William Richard Basham

Renal Diseases

A clinical Guide to their Siagnosis and Treatment

William Richard Basham

Renal Diseases
A clinical Guide to their Siagnosis and Treatment

ISBN/EAN: 9783337140229

Printed in Europe, USA, Canada, Australia, Japan

Cover: Foto ©berggeist007 / pixelio.de

More available books at **www.hansebooks.com**

RENAL DISEASES:

A CLINICAL GUIDE

TO THEIR

DIAGNOSIS AND TREATMENT.

RENAL DISEASES:

A CLINICAL GUIDE

TO THEIR

DIAGNOSIS AND TREATMENT.

BY

W. R. BASHAM, M.D.,

FELLOW OF THE ROYAL COLLEGE OF PHYSICIANS;
SENIOR PHYSICIAN TO THE WESTMINSTER HOSPITAL, AND LECTURER ON MEDICINE,
ETC. ETC.

LONDON:
JOHN CHURCHILL AND SONS, NEW BURLINGTON STREET.
MDCCCLXX.

TO MY

COLLEAGUES OF THE WESTMINSTER HOSPITAL,

PHYSICIANS, SURGEONS, AND LECTURERS,

PRESENT AND PAST,

THIS WORK IS INSCRIBED,

WITH SINCERE RESPECT AND REGARD,

BY

THEIR FELLOW-LABOURER,

W. R. BASHAM.

PREFACE.

CLINICAL teaching, or instruction conveyed at the bedside, directly from the symptoms of the patient, has been long recognised as an integral and most essential part of medical education. But it is only very recently that the clinical examination of the student has been considered as the most effective proof of his practical knowledge of disease; and in all probability before long no student will be granted his diploma to practise his profession until his clinical knowledge of disease has been tested as a part of his final examination.

It is with the view of promoting a practical and clinical knowledge of a class of diseases which are not without their difficulties in diagnosis that the present work has been prepared, with the hope that both student and young practitioner may by it be assisted in their clinical observations.

The author's experience as a Lecturer on Medicine, and Clinical Teacher, leads him to believe that such a work will prove useful in clinical teaching, notwithstanding the many excellent works, in this department of medicine, already in the hands of the profession.

The classification of the several diseases of the kidney adopted in the present work is that which the author has followed for many years in his lectures on medicine. Some classification is necessary for the facility of reference, and this has been formed as closely as possible on a pathological basis.

The third part treats of the several properties of the urine—physical, chemical, and morphological—significant or otherwise of organic or functional disease of the kidneys or other organs.

The author believes that the student who makes himself familiar with the various indications which are to be gathered from a proper examination of the urine will prove as expert in the diagnosis of this class of disease as the proficient in the use of the stethoscope becomes in the diagnosis of diseases of the lungs and heart.

17, CHESTER STREET, BELGRAVE SQUARE;
February, 1870.

CONTENTS.

PART I.

PAGE

GROUP I.—Diseases marked by symptoms more or less of an inflammatory character; excited by various causes, and indicated by bloody, albuminous, or purulent urine; the urinary sediment often exhibiting specific microscopic objects . . 1

NEPHRITIS.—*Inflammation of the Kidneys.*

CAUSES AND PATHOLOGY.

Cause 1. Idiopathic—doubtful . . . 7
  „ 2. External—Injuries, Blows, Kicks, particular substances applied to the Skin . . 8
  „ 3. Internal—substances taken internally, and displaying specific effects on the Kidneys: Turpentine, Cantharidis, Nitre, Copaiba, &c., Lead and Phosphorus, acting remotely . 9
  „ 4. The effect of certain febrile poisons, leading to Congestion and Inflammation of the Kidneys, Scarlet Fever, Diphtheria, Enteric Fever, Small-pox, Erysipelas, Pneumonia . . 15
  „ 5. The agency of Cold and Wet in certain Constitutions inducing Acute morbus Brightii, or Inflammatory Dropsy . . . 22
  „ 6. Nephritis associated with Gout, excited by excess of Uric Acid passing through the Kidneys, giving rise to Gravel, Lithiasis, Nephralgia, and leading to Calculous Nephritis and Calculous Pyelitis . . . 27
  „ 7. The deposit of Tubercle within the Kidney or its outlets, Tubercular Nephritis, Tubercular Pyelitis, Scrofulous Pyelitis, Pyo-Nephritis, Pyo-Nephrosis 36

		PAGE
Cause 8. From Cancer deposited in the Kidneys, Cancerous Nephritis		46
„ 9. Peri-Nephritis		49
„ 10. Nephritis endemic from parasitic Ova		46
„ 11. Nephritis and Pregnancy		48

SYMPTOMS, DIAGNOSIS, AND TREATMENT OF THE ABOVE.

PART II.

GROUP II. — Renal diseases non-inflammatory — without primary symptoms of inflammatory action.—Urine albuminous, rarely purulent. Sediment containing specific microscopic Objects—Casts of the Uriniferous Tubes. Diseases chiefly characterised by evidence of degeneration or decadence of the tubular, cellular, and vascular elements of the Kidneys—fatty, granular, fibrinoid, or amyloid, occasionally associated with, or as the sequel to, obstructed circulation through the Lungs, Heart, or Liver 109

CHRONIC NEPHRITIS, OR CHRONIC ALBUMINURIA.

The four following forms represent the structural changes in this group :—

I. THE GRANULAR CONTRACTED KIDNEY	133
II. THE LARGE GRANULAR FATTY KIDNEY	136
III. THE AMYLOID KIDNEY	137
IV. THE ATROPHIC, CONTRACTED, NODULAR, GOUTY KIDNEY	141

Cause 1. Pre-existing Blood Poison, certain fevers	119
„ 2. The strumous taint	121
„ 3. The syphilitic taint	122
„ 4. The gouty taint	123
„ 5. Certain mineral poisons, Lead and Phosphorus	124
„ 6. Intemperance through alcoholized blood	125
„ 7. Obstructive states of the circulation through the Heart and Lungs	126
„ 8. Amyloid diseases traced to the agency of some pre-existing purulent drain, through some exhausting disease	137
„ 9. Renal cysts, congenital hydro-nephrosis	163

SYMPTOMS, DIAGNOSIS, AND TREATMENT OF THE ABOVE.

PART III.

THE URINE.

	PAGE
Properties of the Urine, physical, chemical, and morphological, significant or otherwise of renal disease	172

Physical Properties—

Colour	172
Odour	177
Aromatic.	
Fœtid.	
Ammoniacal.	
Specific gravity	178
Quantity	179
Frequency of micturition	180

Chemical Properties —

Acidity	181
Neutrality	181
Alkalinity	182
Chylous Urine, Kystein	184
Uric Acid	185
Urates	186
Oxalate of Lime—Oxaluria, its origin and treatment	187
Hippuric Acid	199
Urea, deficient or excessive, methods for estimating	200
The Phosphates, alkaline and earthy, crystalline	204
The fixed Chlorides	209
The Sulphates	211
Alcohol in Urine	211
Albumen, methods for estimating	213
Albuminose, or modified albumen	216
Sugar or Glucose, various tests for detecting	217
Process for estimating the quantity, by volume or weight, of Carbonic Acid produced by fermentation	221
The Polariscope	223
Cystin	224
Leucin and Tyrosin	224
Creatin and Creatinine	225
Bile pigment	225
Cholic Acid, methods for detecting	226

MORPHOLOGICAL CONSTITUENTS OF THE URINE.

Organic forms— PAG
   Mucus 228
   Pus 229
   Blood-corpuscles 230
   Renal casts of the uriniferous tubes . . 231
   The fibrinous blood-cast . . . 231
   The epithelial cast 231
   Pus-casts 232
   Fatty and granular casts . . . 232
   Hyaline casts 233
   Tubercular matter 234
   Cancer cells 234
   Spermatozoa 234
   Vibriones 235
   Sporules 235
   Ova 236
   Echinococci 237

ERRATA.

Page 44, line 9, *for* system *read* symptom.
,, 61, line 35, *for* is *read* are.
,, 63, last line, *for* pictoneum *read* pictonum.
,, 79, last line, *for* calculus *read* calculous.
,, 90, line 24, *for* Miahe *read* Mialhe.

PART I.

NEPHRITIS—INFLAMMATION OF THE KIDNEYS.

Group I.—Diseases marked by symptoms more or less of an inflammatory character; excited by various causes, and indicated by bloody, albuminous, or purulent urine; the urinary sediment often exhibiting specific microscopic objects.

RENAL DISEASES.

PART I.—NEPHRITIS.

CAUSES AND PATHOLOGY.

CHAPTER I.—INTRODUCTORY.

NEPHRITIS.—For clinical purposes, under the term Nephritis several disordered or disturbed states of the kidneys are included which are not in a strictly pathological sense inflammatory. Under this head disorders of the kidneys, excited by various causes, will be included which are indicated by bloody, albuminous, or purulent urine. It must not be forgotten, however, that the urine may have these several qualities, and the kidneys may either not be involved at all, or, if implicated, their state may not be of the character of inflammation; or if inflammatory, the ultimate result will differ in many most important particulars. To the young and inexperienced practitioner these apparent anomalies are, at first, grave obstacles to his forming a correct opinion of the nature of the disorder under which his patient suffers.

In illustration it may be stated that in inflammation of the kidneys, from whatever cause—external violence, blows, poison, cold and wet, calculus, &c.—the urine is scanty, all but suppressed, and bloody. In the inflammatory state of the kidneys after scarlet fever the same conditions are present. In calculus of the kidney the urine may be bloody. But the sequel to, or the effect of, these various causes, both on the kidneys themselves as well as on the constitution of the patient, is very different. In vascular tumours, polypi of the bladder, in prostatic disease, in stone in the bladder, in a particular parasitic affection endemic in certain parts of the world, the urine is bloody and the kidneys free from disease.

The urine is albuminous in many different renal disorders. In both the acute and chronic forms of Bright's disease, in all forms of nephritis, in the gouty or atrophic kidney, in several obstructive conditions of the circulation through the heart, lungs, or liver, in pneumonia, in cholera, the urine is albuminous, so that albuminous urine is equally present in inflammatory as in non-inflammatory diseases of the kidneys. The urine, moreover, may be temporarily albuminous in pregnancy, as well as in some states of disordered digestion. Again, the urine is purulent at the latter state of nephritis proper, but it is also purulent in pyelitis, whether calculous or tubercular; it is purulent in gonorrhœa, in cystitis, in prostatic disease, and oftentimes in stricture. It is manifest, therefore, that neither blood, pus, nor albumen are of themselves diagnostic of inflammation of the kidneys, not necessarily even of disease of those organs, and yet one or other of those qualities is ever present when the kidneys are either inflamed or diseased.

As any of the channels—ureters, bladder, urethra—through which the urine passes in its way from the kidney may be the seat of diseased action independent of the kidneys, blood or pus may become mixed with the urine in its passage outwards, and thus mislead the inexperienced.

Before the microscope was applied to the examination of the urinary sediment, and a greater accuracy obtained in the diagnosis of urinary disease, errors of diagnosis of importance to the welfare of the patient must have been, as we know they were, frequent; now, however, a microscopic examination of the sediment in any sample of disease, compared or taken in conjunction with the symptoms of the patient, will lead with certainty to a correct diagnosis.

It has been already observed that the term nephritis includes many different forms of diseased action in the kidneys. It is only within the limits of the present generation that any distinction has been made between diseases of the substance of the organ and those of the outlets or passages.

The causes of nephritis, or the agents through which inflammation of the kidneys or other disorganizing disease may be established, are various, and the results or effects of these various causes differ materially with the cause. The substance or parenchyma of the organ is affected by some, the outlets and passages by others. The term nephritis, applied indifferently and without distinction to many very different morbid processes in the same organ, has been found inconsistent with the more accurate researches of modern pathology, and a nomenclature is now employed which indicates either the probable cause or the seat, or perhaps both, of the disease in the kidneys. Attention to the differential diagnosis of the various disorders, formerly comprehended under the title of nephritis, will thus lead to a sound, correct, and practical knowledge of these disorders.

It is to Rayer that we must concede the merit of first studying renal diseases with the object of establishing the differential diagnosis between an inflammation of the substance of the kidney and of the conduits or passages leading therefrom. The aid which the microscope affords for correctly estimating the nature and seat of the renal disease, from an examination of the materials passing off with the urine, has laid the foundation for a knowledge of the various morbid conditions to which the kidneys are liable, as correct, as minute, and, perhaps, more easily demonstrated than has been attained by the use of the stethoscope in the diagnosis of diseases of the chest.

The term nephritis, then, without any adjective prefix, will express a more limited range of renal disease than it did formerly. It will be confined to such inflammatory states of the organs as may be excited by external injury, or developed by fevers or the peculiar action of certain poisons taken internally. The term idiopathic nephritis is made obsolete from the belief that there is no difficulty in definitely applying, to each individual case, a prefix which will clearly designate the form, cause, and seat of the renal disorder.

Thus, nephritis simply indicates inflammation of the kidneys, such as may be excited by external violence, wounds, or blows, or similar injuries; or such as may be developed by certain poisons received into the blood, whether absorbed through the skin and received into the system by the breath, or through the agency of particular febrile poisons, such as scarlet fever, erysipelas, smallpox, or enteric fever. Other forms of nephritis should be expressed according to the probable nature or sequel of the disorder. Of the several forms of Bright's disease, one is essentially, in its origin, inflammatory. Rayer calls this albuminous nephritis. The term acute morbus Brightii, or acute albuminuria, is better. Gout and rheumatism are occasional causes of renal inflammation; these, therefore, are designated gouty and rheumatic nephritis. Stone in the kidneys develops inflammatory symptoms in these organs; calculous nephritis is therefore an appropriate prefix, as indicating the cause of the renal disturbance. Tubercle and scrofula may affect the kidneys; tubercular and scrofulous nephritis are therefore appropriate terms. Cancer may be deposited in these organs, and give rise to symptoms, to which cancerous nephritis is the proper and intelligible prefix.

When the pelvis and calyces of these organs are the seat of inflammatory action, independently and apart from, or in conjunction with, disease of the renal substance, the term pyelitis is used to express the distinction. But as the causes of diseased action in the pelvis and calyces are various, the necessity for an adjective prefix distinguishing the variety becomes as necessary as in disease of the substance. The chief and most important forms of pyelitis are the calculous and the tubercular, or scrofulous. Stricture of the urethra, involving attacks of retention of urine, is sometimes the parent of wide-spread mischief, extending from the bladder to the ureters, infundibulum, pelvis, and calyces of the kidney, giving rise to symptoms and changes of structure which will be described hereafter, a condition sometimes called pyelo-nephritis.

Rayer and other French pathologists have described a series of morbid conditions, inflammatory chiefly, affecting the tissues surrounding the kidneys, but always occurring as the sequel to, or the effect of, some antecedent disease of the kidneys, and to which they have applied the term *peri-nephritis*. In contusions of the loins, in injuries or wounds of the back, inflammatory action around the kidneys may be set up. In some fevers, exceptionally, abscesses have formed external to the kidneys. In calculous disease of the kidney, in tubercular disease as well, where a renal abscess exists, the purulent collection may ulcerate its way outward through the loins, and will, in these several cases, excite inflammatory action in contiguous and surrounding textures, and to such conditions the term peri-nephritis has been applied. It appears undesirable to multiply the names of diseases and to apply to exceptional and rare occurrences, which are always secondary, a term which would imply that the disease was primary. Reference will hereafter be made to the character of the symptoms occasionally excited by renal disease thus extending beyond the limits of the organs, but it is not desirable that peri-nephritis should be treated as a distinct or substantive disease.

It may readily be conceived from these observations that the various forms of nephritis will give rise to symptoms to a certain extent differing one from the other, but in these very differences will be found the key to a correct diagnosis.

There are, however, a few prominent symptoms which may be said to be common to all varieties, and, therefore, pathognomonic of the early stage of nephritis, from whatever cause. These are scanty and all but suppressed, high-coloured, turbid, or bloody urine; frequent, difficult, or painful micturition; urgent pain in the loins, sacrum, pubis, perinæum, or urethra, with more or less febrile disturbance and vital depression.

If we select a typical case of nephritis from any, whatever, cause, it will be found that these symptoms are present in greater or less intensity in each; but it is to the prominence of one or other of these symptoms beyond the others which, in the early stage, often guides the physician to a correct diagnosis of the particular form of nephritis which the patient suffers. Thus, in nephritis from local or external injury the all but suppressed urine, which appears little more than pure blood, the difficult and painful micturition, and the urgent pain at the neck of the bladder, are far more prominent symptoms than the accompanying febrile disturbance; while in the nephritis of acute albuminuria the rigors, prostration, irritable state of stomach, anorexia, heat of skin, and accelerated pulse, are more prominent and noticeable symptoms than those which have a more direct reference to the urine or the kidneys. The urine is scanty, high coloured, and may contain blood, there is also frequency of micturition, but these symptoms are not the leading or the more prominent ones. It is, therefore, by carefully estimating the force and character of both the general and local symptoms that a correct diagnosis of the special form of nephritis is arrived at. It will be conducive, therefore, to this object, to describe in succession, and in the order in which the causes have been named, the symptoms of the several varieties of nephritis, laying emphasis on those symptoms which may be considered pathognomonic of each form.

CHAPTER II.—CAUSES AND PATHOLOGY.

Inflammation of the Kidneys, with Bloody, Albuminous, or Purulent Urine.

THE causes which may excite inflammation of the kidneys, in whatever form, may be conveniently divided into—1. Idiopathic—doubtful. 2. Those which act externally or from without. 3. Those which operate from within the organism, either developed in the kidneys in the course of other diseases, or originating in the kidneys themselves from retention of some excrementitious material within them.

§ 1. NEPHRITIS—*Idiopathic;—from external injury;—from poisons, either external or internal.*—I have elsewhere* expressed a doubt if nephritis ever exists as a primary or what is usually called an idiopathic disease. Whether inflammation of the kidneys can arise apart from the influence of some pre-existing internal cause, such as the irritation of calculus or gravel, the presence of tubercle, scrofula, or cancer—the influence of certain febrile poisons, scarlet fever, enteric fever, &c.—the agency of certain exciting causes, in predisposed constitutions, bringing on a particular form of inflammatory action, followed by dropsy and not by the formation of pus—in fine, whether the kidneys are susceptible of acute parenchymatous inflammation apart from any of the above-named exciting causes, as the lungs are in pneumonia. Cases are occasionally met with in which all the characteristic signs of nephritis are present, followed by an abundant formation of pus discharged with the urine, which may easily be mistaken for tubercular or scrofulous pyelitis, but which the sequel of the case forbids us to

---

* 'On Dropsy,' 3rd edition, p. 414.

think could have originated in such deposit in the kidneys, as it is contrary to all experience to find a cure in such forms of renal disease. That such cases are occasionally met with is admitted, and reference, under the head of Symptoms and Diagnosis, will be made to two which have fallen under the care of the author.

Causes which act externally in exciting inflammation of the kidneys.—These are penetrating wounds or injuries to the loins, contusions by kicks or blows, or hurts by bodily straining, such as wrestling, jumping, or any violent wrench or shock after the manner of concussion.

The action of certain medicinal substances applied to the skin and exercising specific effects on the kidneys might be included among the causes acting from without. But though their application is external, their action is through the same channels as if taken by the mouth; they pass into the circulation by absorption, and develop their effects while passing through the kidneys.

The action of turpentine, cantharides, and similar agents, will be included in the second group.

§ 2. NEPHRITIS *from external injury.*—Cushioned as they are by the surrounding textures and protected by their situation, injury to the kidneys by external violence is comparatively rare. Accidents by machinery and the crushing injuries caused by railway collision excepted, the kidneys are not often the seat of inflammation through external agency. Nevertheless cases are from time to time observed.

The primary effect of external injury, whether by blow, kick, wounds, crushing or penetrating, will be inflammation; that is, an intense inflammatory congestion, with blood effused into the urinary tubes from rupture of the Malpighian or other vessels. A violent blow, such as a kick or a violent shock, may rupture a number of the Malpighian bodies, and blood would be poured at once into the tubules, and thence run through the ordinary urinary outlet—the ureters. A diffuse inflammatory action throughout the entire organ will follow. In the case of crushing accidents or penetrating wounds directly lacerating the organ, the hæmorrhage would necessarily be freer and more abundant, pouring, from the lacerated branches of the renal artery as well as the wounded Malpighian bodies, blood freely and directly into the pelvis of the kidney. Inflammation follows, the secretion of urine is suspended, so that, in addition to an abundant hæmorrhage, there will be a suppression of the renal secretion. The ordinary products of inflammation follow, pus ultimately appears in the place of the blood, and in favorable cases a perfect and permanent cure results.

See further under the head of Symptoms.

§ 3. NEPHRITIS, *causes*.— (*a*) *External*. The effect of certain substances applied superficially to the skin.

(*b*) *Internal*. The effect of the same substances taken internally, and acting specifically on the kidneys—turpentine, cantharis, copaiba, and other oleo-resins, nitre, lead, phosphorus, &c.

The effect of these agents, whether applied to the skin or taken internally by the mouth, may be described under the same section, as their effects are the same, and the channel, the blood, through which they operate, similar.

Turpentine, when applied to the skin, is quickly absorbed, and in the course of a very short time imparts to the urine a peculiar violaceous odour. Its presence in the blood stimulates the action of the capillaries, and hastens the flow of blood through them. This effect is very apparent on the kidneys, as an increased flow of urine is the immediate consequence. It is this action on the capillaries which constitutes the therapeutic value of turpentine applied externally to the chest in inflammatory states of the lungs. Ordinarily, the effect of turpentine to the skin on the kidneys is simply as an excitant to increased urinary secretion, and it is only in exceptional cases that its action passes beyond this and induces engorgement or even inflammatory action. But these exceptional cases are of great importance in treating of diseases of the kidneys. In the progress of cases of acute, as well as of chronic albuminuria, inflammatory complications are well known suddenly to spring up. Among these, symptoms of inflammatory engorgement of the lungs are, perhaps, the most frequent, and it is in these cases that turpentine, applied externally to the skin, acts so prejudicially on the kidneys. The function of these organs is already embarrassed. In the acute form of morbus Brightii the kidneys, in the early stage, are in a condition of inflammatory engorgement. The action of turpentine in such cases induces still further embarrassment, and may lead to hæmaturia and suppression of urine. The use, therefore, of turpentine as a local epithem in pulmonary inflammation in Bright's disease must be carefully avoided. I do not recollect that any other evil effects have followed the application of turpentine to the skin. Taken internally, the action of turpentine on the kidneys is more obvious; it has been employed in various disorders, both in small and large doses. In respect to quantity, there is something very remarkable in its action. In large doses, from half an ounce to two or three even, its action has been noted chiefly as purgative and stimulant, producing effects not unlike an equal quantity of gin. It is regarded as a powerful anthelmintic in tape-worm, and in the doses usually prescribed for that purpose its action on the kidneys, beyond impregnating the urine with its peculiar aroma, is not obvious. It is, in small doses, from twenty minims to half a drachm, employed chiefly in disorders of the mucous membranes, such as gleet, leucorrhœa, &c.; in somewhat larger doses, from one to two drachms, it has been given in sciatica, lumbago, and some nervous disorders. It is in these quantities that its deleterious effects have been noted. These are a sense of great distress from ineffectual efforts to pass urine, the desire to pass which, accompanied by a heavy aching pain at the pubis and neck of the bladder, comes an hour or two after the turpentine has been taken; sometimes two or three doses may be taken before the renal effects are observed. The first drops of urine passed are bloody, and the suppression of urine may continue for some hours, the secretion slowly returning, with abatement of all the more distressing symptoms. If the turpentine be immediately discontinued, and the patient remains perfectly quiet and recumbent, and drinks moderately of diluents, in a few hours all disagreeable effects disappear. If there be any constitutional predisposition to renal disease, the renal engorgement, with bloody and suppressed urine, will last longer. Warm baths may be necessary, but it rarely happens that local depletion by cupping is required. · In employing turpentine, therefore, as a therapeutic agent, whether externally or internally, care should always be previously taken to determine the state of the renal organs.

The active principle of the Spanish fly, *Cantharis vesicatoria*, exercises specific effects on the kidneys. These are rarely observed to follow its outward application in the well-known form of a blistering application. But when taken internally, hæmaturia, suppression of urine, and even bloody stools, have followed its administration. It is a very uncertain remedy for good; it is a very energetic one for evil. In certain disorders of the mucous membranes of the genito-urinary organs, in incontinence of urine, and in some other conditions of these parts conveniently, for the want of better knowledge, attributed to debility, this energetic poison, chiefly in the form of tincture and in doses of from five to twenty minims, has been employed. It has received a spurious reputation for the cure of gleet, gonorrhœa, and leucorrhœa; it has also been thought to possess emmenagogue properties. Its action on the genital organs of either male or female is entirely through its violent specific action on the intestines and kidneys. It produces a bloody flux from the first, and bloody discharges with suppression of urine from the second. It has been basely and criminally used under the notion of exciting venereal desire. It acts in no such channel; any amount of excitation on the genital organs are only a part of its poisonous agency and proof of its deadly effect on the unfortunate victim of its administration. Medicinally employed, therefore, it can never with certainty be predicted that the patient may not become the subject of nephritis, with all its miserable consequences. Unlike turpentine, the deleterious effects of Spanish fly on the kidneys do not disappear with the discontinuance of the agent.

After the nephritic symptoms have become apparent very active means are requisite for their management. Soothing oleaginous purgatives, with opium, the warm bath, absolute rest, mucilaginous drinks, milk and water, and mild vegetable diet, and abstinence from stimulants, are requisite. The urine for many days, even after the secretion has been re-established, contains microscopic evidence of blood, and in one case which came under my observation some years since, in a young woman of nineteen, suffering from amenorrhœa, in which half-drachm doses of the tincture had been prescribed with the notion of recalling the absent uterine function, and in whom retching and vomiting had occurred with suppression of urine, followed by hæmaturia, the urine contained microscopic evidence of blood-corpuscles three weeks afterwards, and on several occasions blood-casts of the large straight tubes were seen. The urine contained a small quantity of albumen. It was near ten weeks before this patient could be pronounced free from the effects of the cantharidin on the kidneys. The effect of turpentine and Spanish fly on the kidneys is first to excite a state of engorgement of the tubes, and subsequently, as the congestion passes away, a catarrhal condition of the straight tubes of the pyramids.

In the 'British Medical Journal' for June 26th, 1869, there are some remarks by Dr. Edward Mackey, Lecturer on Materia Medica in Queen's College, Birmingham, on the value of tincture of cantharides in some forms of pyelitis. The cases recorded appear to have been calculous pyelitis in two women, and the progress differed in no respect from similar cases the treatment of which has been by other remedies than cantharides. The subject of calculous pyelitis will be treated hereafter.

Nitrate of potash in concentrated dose has been said to produce strangury and bloody urine. An ounce or more, in no greater quantity of water than may be necessary for its solution, has been taken in mistake for other salines—Epsom or Glauber's salts, for instance. These cases are very rare, and, therefore, it may be here only necessary to record the fact that this salt in a concentrated form is capable of exciting nephritis, while, on the other hand, equal quantities, largely diluted, may be taken with advantage, and with no other effect on the kidneys than a gentle stimulus to increased secretion of urine, that fluid being strongly impregnated—in fact, continuing to excrete the nitrate so long as any remains in the blood. In the experience of the author this salt, largely diluted, an ounce to a quart of water, acidulated slightly with lemon juice, and flavoured with the peel, is an invaluable and reliable remedy in acute rheumatic fever.*

The action of turpentine and cantharides on the kidneys is direct and immediate; that of some other poisonous substances, such as lead and phosphorus, indirect, slower, and more insiduous, and leading eventually to fatal degeneration of the cellular element of the kidneys.

Phosphorus, as well as the salts of lead, are largely employed in certain manufactures. They are both poisonous, and may find their way into the organism through the carelessness or neglect of cleanliness on the part of the workmen.

The effects of lead on the system are familiar in the form of painters' colic and dropped or paralysed wrists; those of phosphorus are less known, chiefly, perhaps, because it is only of late years, within forty certainly, that phosphorus has entered so largely into the requirements of the age, in the manufacture of the varieties of matches for procuring light, a valuable exchange for the old, tedious, and uncertain sulphur match and tinder box.

Neither the salts of lead nor phosphorus appear to produce nephritis or any active diseased state of the kidney. They severally lead to slow changes of structure, in which degeneration of the renal epithelial cells is the most obvious, accompanied by albuminous urine.

* See a paper in the 'Medico-Chirurgical Transactions,' vol. xxxii, 1849, "On the Nitrate of Potash in Acute Rheumatism," by the author.

The influence of lead as a predisposing cause of gout, however, must not be forgotten. Dr. Garrod* has made it the subject of special inquiry. At first it might be suspected that the habits of these artisans were more likely to induce gout than the poisonous influence of the lead with which they worked, but Dr. Garrod has clearly established the fact that gout is more prevalent among those whose system has been impregnated with lead than among other workmen earning equal wages, and whose habits of indulgence were similar. I have had many opportunities of bearing witness to this fact, which is admitted now by every physician of experience.

It is worthy of remark that in many cases of painters' colic the urine is albuminous, but without dropsical effusion; it is also very certain that in a great number of similar cases the urine is free from albumen.

In the former class the kidneys are in a progressive stage of atrophy. Can we, then, regard lead as exercising any specific effect on the kidneys? I think not, except so far as susceptibility to gout does so. In every case of albuminous urine occurring in painters that I have seen there has been evidence of gouty attacks. It is only when the constitution of the patient is broken down by the long-continued influence of irregular and dissipated habits, conjoined with the poisonous effects of the lead and gout, that an albuminous state of the urine declares the general decadence of cell structures throughout the body—a degeneration in which the renal cells, of course, participate. In fatal cases the kidneys have been found granular, the cortical surface exhibiting a sabulous appearance, the renal cells and tubes being granular and fatty; while in others who had suffered from gout the kidneys were in a progressive state of atrophy, the cortical substance being reduced to a very narrow line.

Similar remarks apply to the action of phosphorus. Those whose employment causes them to handle this elementary substance suffer dreadfully from its poisonous effects—frightfully

---

* 'Gout and Rheumatic Gout,' p. 270.

rapid erosion of the gums, decay of the teeth and alveoli, ulcerations of the mouth and cheeks, and in fatal cases marked evidence of extensive renal degeneration.

Dr. Habershon has reported a case of the effects of phosphorus on the kidneys.* During life the urine contained albumen, and under the microscope a large quantity of epithelium, with casts of tubes and a few blood-corpuscles, were observed.

After death the kidneys appeared of a light yellowish-pink colour, the cortex injected, the pyramids of a deep red, and the uriniferous tubes loaded with fat. This extreme fatty degeneration of the epithelium of the tube is the most noticeable feature in all the cases recorded.†

§ 4. NEPHRITIS, *a sequel to scarlet and other fevers.*— It is a matter of observation that in some epidemics of scarlet fever the proportion of cases of renal dropsy appear to be greater than in others. This has been accounted for variously. Some observers have attributed the concurrence to the severity of the epidemic. Others have thought that the primary rash was defective, and the febrile virus not perfectly eliminated by the skin, the kidneys being left to complete the process of depuration. Some have conceived that, during the period of cutaneous desquamation, the surface of the body being then specially susceptible to the influence of temperature, a chill then felt would quickly lead to disturbance in the equilibrium of the circulation, and the kidneys, so sympathetic in their action with the skin, become congested and inflamed.

I have long been of opinion that neither of these methods of interpretation is the correct answer to the question why the kidneys are so prone to become disordered after scarlet

* 'British and Foreign Medical Review,' April, 1868.

† See two papers in the 'Medico-Chirurgical Transactions,' vol. l, 1868, "On Acute Poisoning by Phosphorus," one by Dr. Habershon, and the other by the late Dr. Hillier on poisoning by the same substance; in both papers fatty degeneration of the renal cells were among the post-mortem appearances. Jaundice appeared to be common to these cases, and albuminuria coincident with jaundice.

fever. It appears to me that some antecedent bodily condition must exercise a large influence in predisposing the renal organs thus specially to suffer in a certain number of cases of scarlet fever. Observation convinces me that in those families in which there is an obvious strumous taint these renal disorders after scarlet fever are more frequent than among those free from this sign of deteriorated vital energy.

I do not presume to say that this can be proved, but I am firmly of opinion, justified by long observation, that the antecedent fever is the lesser factor in this proposition.

I have elsewhere expressed a similar opinion.

The renal disorder after scarlet fever commences as a true inflammatory engorgement. Death taking place in the first week, the kidneys are found of various degrees of inflammatory redness, purplish-red, brownish-red, chocolate-red; the cortex covered with arborescent vessels tinged with blood; the Malpighian bodies swollen, many ruptured, and blood escaping into the uriniferous tubes, mixed with the scanty and all but suppressed urine. The microscope exhibits these tubes filled with blood, and it is very easy by pressure to squeeze from the apex of the cones a fluid which is blood mixed with fibrinous casts of these tubes, accompanied by granular matter and renal epithelium. When the patient survives this period, and the congested state of the kidneys subsides, this renal epithelium is cast off, for the inflammatory stasis has so injured the vitality and secreting power of many of these cells that they have become effete and useless. This destruction of the renal cells has been called a process of desquamation, and it has been supposed that the uneliminated part of the scarlatinal poison falling on the kidneys has been followed by a shedding of these cells, just as the cuticle is shed or peels off, and by analogy the disease has been designated desquamative nephritis.

The name signifies little, provided the pathological process be understood. But it must be clearly understood that this casting off of the renal cells in the form of granular epithelial casts is not a process similar to what takes place in the skin.

The textures are anatomically different, and the conditions not equal.

The epidermic cells of the cutis are protective; the epithelial cells of the kidneys are secreting. The cuticular structure is built up of a succession of layers of cells in different stages of development, the oldest and most external and superficial being constantly shed and renewed by the progressive development of those beneath. This process of desquamation of the matured and outermost cells goes on constantly through life, and forms one of the most distinctive properties of the epidermic layers of the skin.

The process in some cutaneous disorders becomes exaggerated, and the outer layers are shed faster than the new cells can be renewed; the young immature cells being an indifferent protection to the tender and vascular cutis. Such frequently happens to portions of the skin in the free desquamation which follows scarlet fever.

In the true epithelial tissues a similar process of replacement of the old and effete cells by younger and more active elements also exists.

The epithelial layer of the renal tubes, however, is a single layer, arising direct from a basement or germinal membrane, from which these cells are developed. Moreover, these renal cells are secreting cells, and their process of renewal and replacement is much slower and less frequent than in the epidermic series. There is every reason to believe that the renal cell follows the law of other epithelial structures, and when effete and useless is thrown off and replaced by younger cells.*
But in health there is no evidence of this, as no *débris* of cell structure is ever seen in healthy urine.

But let the equilibrium of the circulation within the kidneys be disturbed sufficiently to induce a stasis in the renal capillaries, the nutrition of these cells is quickly disturbed, and they rapidly become defunct, and are cast off in abundance as epithelial casts of the renal tubes. This destruction of renal secreting structure is to a great extent proportioned not

---

* Carpenter's 'Physiology,' edited by H. Power, pp. 342, 343.

so much to the intensity of the inflammatory engorgement, as to its duration.

Those cases in which the kidneys are relieved in the early stage by a free unrestrained hæmorrhage, recovering quicker and more favourably than those in which a smoky or greenish discoloured urine, indicates a lesser degree of engorgement, but one of a more serious and durable kind.

This casting off in abundance of the defunct or deteriorated cell is followed by a marked alteration in the character of those which follow, so that, unlike what takes place in the desquamation of the epidermic layer in scarlet fever, the younger cells following in succession being complete and perfect, those renal cells which follow or are developed in the place of those cast off are degenerate and inefficient, and are as quickly as formed detached and thrown off also. A daily or frequent microscopic examination of the sediment in a case of scarlatinal dropsy will demonstrate this. The renal cells at first thrown off are opaque and the nucleus obscure, and they appear moulded together in the tubes from which they are detached, and are more or less granular. These are soon accompanied by cells more manifestly departing from the type of the renal epithelial secreting cell. The cell is larger, the nucleus, no longer simple, may be reniform or even multinuclear; the contents of the cell become distinctly coarsely granular and fat-granules make their appearance. These varieties, called severally exudation-corpuscles, sometimes mucous corpuscles, sometimes inflammatory or Gluge's corpuscles, or compound granule-cells, are in reality nothing less than crops of abortive cells developed from the germinal membrane, and effete and useless to supply the place of the first cells which the disordered state of the kidney had caused to be ejected. This is the process established by the engorgement of the kidneys which may succeed to an attack of scarlet fever. The same process in every respect similar is observed to take place in acute morbus Brightii—the same inflammatory stasis, the same appearance of blood-casts, of fibrinous casts, epithelial granular casts, followed by corpuscles or cells of the abortive characters just enumerated. So long as these objects are seen in the urine, so long is the process of degeneration proceeding. The renewal of efficient renal cell-structure is marked by the disappearance of the inferior type of cell and the gradual subsidence and eventual absence of albumen from the urine. In fatal cases the process of deterioration of cell development goes on with an accelerated force; the granular condition of the abortive cells becomes greater; the renal tubes become choked and blocked up, partly by effete granular cells and partly by the granular material derived from the disintegration of these cells. These abortive cells rapidly break up if not quickly washed out and carried off by the current of the urine. The distension of the tubes is thus greatly increased, added to which the intertubular structures become infiltrated with granular exudation, the obstruction to the circulation from these accumulated morbid products, both within and external to the renal tubes, reaches a point which now causes the organ to become pale and anæmic, except the cones, which are sometimes of a delicate salmon hue, and others more deeply injected, while it is largely increased both in volume and weight. The cortical surface though pale is studded with star-like radiations of capillary vessels, and when the capsule is removed the surface is seen studded with granulations varying in size. Such kidneys may weigh eight, ten, or even sixteen ounces each. Such are the pathological changes which the kidneys undergo in the disorder which may follow scarlet fever.

More rarely the course of other fevers is marked in the sequel by renal disturbance. Diphtheria may be considered the most nearly akin to scarlet fever in the susceptibility of patients to renal disturbance. In fatal cases the kidneys have been found in a state identical with what is seen after scarlet fever.

In the 'Pathological Transactions,' vol. x, p. 317, is an account of the post-mortem appearances of the kidneys in this disease, which since that period has conclusively settled the point of the identity of the pathological changes in both diseases. The reporters are Dr. George Johnson and Dr.

Wilks, who sum up the account of the microscopic appearance of the kidneys as follows:

"It will be seen, then, that the condition of this kidney, as observed both by the naked eye and by the aid of the microscope, is identical with that which is found in a large proportion of the instances of acute inflammatory renal disease which occur in connection with scarlatina; and we know of no means by which an inflamed kidney from a patient who has died of diphtheria can be distinguished from the analogous affection which occurs so commonly in scarlatina and less frequently in connection with various other zymotic diseases."

In tracing any case of albuminuria to its probable origin this fact of the analogous effects of diphtheria in causing renal disturbance with those observed to follow scarlet fever must not be overlooked. I have seen many cases of persistent albuminous urine after dipththeria, continuing as many as two or three years, without any disturbance of the general health, apart from the anxiety naturally attendant on the knowledge of this abnormal state of the urine.

Albuminous states of the urine have been observed in erysipelas, in measles, smallpox and typhoid fever, also always in the reactionary stage of cholera. Occasionally and exceptionally the kidneys become affected in the progress of other fevers. Enteric fever, smallpox, and measles, sometimes afford examples of this renal disturbance. A case occurred to Dr. Ogle at St. George's Hospital, mentioned in the 'Lancet' of July 31st, 1869, p. 158, of enteric fever with hæmaturia. A girl of seventeen had fever with symptoms of an hysterical character; the urine during the progress of the fever was observed to be dark, thick, and opaque, and contained a large quantity of broken-down fibrinous material, renal epithelial cells, granular casts of the uriniferous tubes and free blood-corpuscles; it was highly albuminous. The deposit in the urine gradually cleared off, and this secretion on her discharge was in all respects healthy. Similar conditions with like results have been observed by myself in a case of smallpox.

Chronic renal disorder is more frequent after erysipelas, enteric fever, measles, and smallpox, than the acute. This will be treated more particularly under the section of chronic albuminuria.

In pneumonia of exceptional severity the urine becomes albuminous. In this disease the kidneys become congested (passively) from obstruction to the circulation through them, in consequence of the retardation of blood in the cava and right side of the heart. This congested state of the kidney passes away with the restoration of the pulmonary circulation, and with it the albuminous state of the urine. In cholera the urine is found albuminous when that secretion slowly returns during the period of the secondary febrile reaction. In two cases which I watched through the period of convalescence, in the epidemic of 1848, the albumen daily decreased till the patient had recovered his ordinary vigour. In a fortnight no trace could be found.

In these several diseases the kidneys do not suffer that form of acute inflammatory action which characterises the sequel of scarlet fever and of diphtheria.

There are also certain diseases of the heart, lungs, and liver, characterised by obstructive conditions of the circulation through them, in which the urine becomes albuminous. Thus in mitral diseases, in pulmonary emphysema, as well as in cirrhosis of the liver, the current of the circulation is impeded, and serous effusions in the shape of dropsy make their appearance. The urine eventually becomes albuminous, but this condition depends on the degree of impediment to the circulation, not on any inflammatory or other primary disorder in the kidneys.. The obstructed circulation through the vena cava inferior retards the flow of blood through the kidneys; venous congestion results, and the effect is equivalent to a ligature round the emulgent vessels of the kidneys, which Dr. Robinson, of Newcastle, demonstrated as effective in inducing albuminous urine.*

This amount of obstruction to the circulation through the

* 'Medico-Chirurgical Transactions,' vol. xxvi.

kidneys is measured by the amount of retardation of the blood in the cava, and consequently the albuminous state of the urine depends on that retardation and not on any primary disorder in those organs. It thus happens that in the early stages of valvular disease of the heart, as well as in emphysema, the urine is at first, and indeed, in some cases, throughout the whole disorder, free from albumen. In most cases of ascites from hepatic obstructive disease the urine is also free from albumen until the pressure of the accumulated fluid in the abdominal cavity becomes so great that its effect is felt on the emulgent veins of the kidneys.

It must be clearly understood then that albuminous urine accompanying these forms of disease is at the early period of its appearance significant of no other condition of the kidney than that of venous congestion, and the albumen is derived probably from the infiltration into the renal tubes of the serosity of the blood just as the areolar tissue becomes infiltrated with serum, causing the accompanying dropsy. In these cases of cardiac and pulmonary dropsy the kidneys may, however, ultimately become the seat of organic change. The congestion of the renal plexus of veins surrounding the tubules when long existing eventually interferes with the due nutrition of the organ, and degeneration of the renal cells with accumulation of granular and fatty material within them testifies to the effect which long-continued derangement in the circulation of an organ has upon its healthy condition.

These forms of renal disorder will be further considered under the head of chronic albuminuria from heart or lung disease.

§ 5. NEPHRITIS—*Acute morbus Brightii; acute albuminuria; inflammatory dropsy.*—Effects of exposure to cold and wet on certain predisposed constitutions.

Pregnant with ill-consequences, not alone to the kidneys but to other vital organs, and jeopardising the life of the individual, is the effect of exposure to cold and wet in certain constitutions, particularly the strumous.

An antecedent predisposing state of the body seems necessary to concentrate the ill-consequences on the kidneys, or the effects of such exposure would be more uniformly alike; whereas it is well known that what acts prejudicially on one is innocuous on others. Irregular habits of life, an alcoholized state of the blood, the scrofulous temperament, particular states of health, the sequel of some fevers, may be mentioned as among the most frequent predisposing causes.

Thus, a working-man gets drunk, sleeps off his debauch in the open air; gets chilled; becomes feverish; and in a day or two has scanty urine, charged perhaps with blood, with œdema of the face, and subsequently general anasarca. Another may be of temperate or moderate habits, who for hours, in his calling, is exposed to wet, and may have suffered like exposure so often as to regard it with indifference, suffers wandering pains with rigors, thirst and feverishness, passes but little urine, and that dark coloured, even bloody, whose face becomes puffy, followed by dyspnœa and general anasarca.

These are examples of the operation of cold and wet in inducing inflammation of the kidneys, a form of disease identical with the inflammatory dropsy of older observers, and of acute albuminuria or acute morbus Brightii of more recent days.

The post-mortem appearances of the kidneys in acute morbus Brightii will vary somewhat according to the period of the disease at which life terminates. So many complications of an inflammatory type may suddenly arise at any period in the course of this disease, that the patient may be cut off at the outset within the first week or two, or some weeks may pass and pericardial effusion may suddenly terminate life, or in more protracted cases bronchial disorder, with œdema pulmonum or convulsions, or cerebral attacks indicative of uræmic poisoning, may severally bring the case to a close. The post-mortem conditions of the kidneys thus will vary in relation partly to the duration and partly to the intensity of the disorder. If death takes place in the earliest stage of an attack of acute morbus Brightii the kidneys present all the characteristic appearances of true nephritis. They are swollen, tinged with blood, which on section flows freely, their colour varies from a purple red to a chocolate brown. The cortex is dark and mottled, the cones of a deep maroon red, and striated with darkened lines. The Malpighian bodies are swollen and the ruptured vessels empty themselves into the renal tubes, which are filled with blood and fibrinous coagula. The prolongation of life for a few weeks effects a notable change in the appearance of the kidneys. The colour is no longer of the dark chocolate character—it has more of a yellowish hue; spots of injection are seen here and there, and vascular ramifications on the surface. The cones are of a deep red hue, and contrast much with the mottled yellowish aspect of the cortex; they are strongly striated with red lines. The cortical surface is sabulous or granular, displaying the well-known granulations of Dr. Bright. The kidney is increased somewhat in weight. By pressing the cones a turbid fluid may be squeezed out, which, by the aid of the microscope, will be found to contain epithelial granular casts, fibrinous granular casts, blood casts, and various forms of abortive cells.

A few weeks' prolongation of life, and the kidneys will exhibit the appearance already described as characteristic of the kidney in renal dropsy after scarlet fever; increase of volume and weight; the cortical surface pale and anæmic, with here and there some stellar vascularity. On removing the tunic the surface is smooth, and a section displays the internal aspect of the cortical part greatly increased in thickness, and appearing to make inroads between the pyramids, which appear narrowed and limited; the colour throughout pale and wax-like, except that the cones are of a delicate pink or salmon colour. A milky fluid can be pressed from the cones, and, when examined by the microscope, casts, epithelial, granular, and hyaline, casts loaded with abortive cells, with highly resplendent fat-granules, isolated and in groups; and exudation corpuscles loaded with fatty material, and coarsely granular cells of abnormal size are the objects seen. The material which adds volume to the entire organ,—the kidneys often weighing ten, twelve, even sixteen ounces each, which thickens to twice its breadth the cortical substance, and which ramifies between the cones is, by microscopic examination, a fine granular material, rich in fatty or highly resplendent nuclei, and appears to be the result of an energetic tumultuous inflammatory action, an exudation in fact from an anomalous condition of the blood.*

Such are chiefly the post-mortem states of the kidney in the acute form of Bright's disease.

The subject will be further discussed in the part devoted to renal diseases of non-inflammatory origin, under the head of chronic Bright's disease.

§ 6. RHEUMATIC NEPHRITIS.—This term originated with Rayer, who, observing isolated patches of inflammatory action, in the substance of the kidneys, both in the cortical as well as in the pyramidal portion of the organ, in fatal cases of acute rheumatic fever, attributed the renal disease to the same cause which excited inflammation in the joints, or heart, and gave to this morbid condition of the kidney the name of rheumatic nephritis.

Considering the frequency of acute rheumatism, and the number of cases which annually come under treatment in the metropolitan hospitals, the occurrence of renal disturbance during life is unquestionably rare : and this is the more remarkable when it is recollected how loaded the urine is with the products of excessive nitrogenous metamorphosis, how high the specific gravity, 1028—36, arising from the great excess of urea, urates, and uric acid, urine pigment, and extractive; the very high specific gravity is partly due to the deficiency of water, arising from the excessive sweating, but also to the large increase in the absolute quantity of the above-named urinary ingredients. Functionally thus taxed to the utmost, it is remarkable how rare, during life, in this disease, the kidneys exhibit any symptoms of organic disturbance, or any approaching to those of inflammatory engorgement.

However, in fatal cases of acute rheumatic fever, when cardiac complications have existed during life, appearances have

* See 'Rokitansky,' vol. ii, p. 8, *et seq.*

been observed in the post-mortem examination of the kidneys identical with those recorded by Rayer, but now differently interpreted. This state of the kidney is now known to arise from embolism, or blocking up of the smaller vessels of the kidneys by the transport of fibrinous coagula from the valves of the heart, or by the occlusion of the renal intertubular vessels by fibrine spontaneously exuded within them. These transported coagula, as well as those formed within the renal vessels, serve to arrest the circulation within the organ at these points. Around these fibrinous blocks a zone of inflammatory redness is apparent, so that the kidneys, in a recent specimen appear studded with red spots; or, if the life of the patient be prolonged a few days after the coagula have formed, minute points may be observed elevated above the surrounding parts distinguishing the spots of inflammatory action.

Dr. George Johnson, who has contributed so much to clear many obscure points in renal pathology, however, doubts whether the views of Virchow, Kirkes, and others, are tenable. Whether, as they think, fibrinous coagula from the valves of the heart can, in being arrested in the minute vessels of the kidneys, produce the appearances described. In a valuable contribution to the Pathological Society, in 1857,* "On the Minute Anatomy of the so-called Fibrinous Deposits in the Kidneys," he is of opinion that the morbid process in the kidney begins with the formation of coagula in the intertubular capillaries. In consequence of the blood stagnating there, the Malpighian capillaries become gorged, which explains the intensely red margin of the recent deposit. He thinks this could not be so if the fibrinous coagula passed through the renal artery, and plugged the small arteries and the Malpighian vessels first. The immediate effect of which would be the cutting off of the supply of blood to the tube.† The pathology, therefore, of this disease of the kidney, attending on acute rheumatism, is of the nature of the primary disorder, an exudation of fibrine within the canal of the intertubular vessels, such as occurs on the valves of the heart, and

* 'Transactions,' vol. ix, p. 305—6. † Ibid., p. 305.

a consequent obstruction to the current of blood, with ulterior changes, if life is prolonged, both in the exudation itself as well as the epithelium of the urinary tubes of the usual fatty transformation.

While I freely endorse these observations of Dr. G. Johnson, having on several occasions been able to verify them, I nevertheless have been able to demonstrate to pupils the correctness of the views of Virchow and Kirkes, that fibrinous coagula detached from the valves of the heart are arrested in the renal artery.

Moreover, the records of the Pathological Society establish this fact beyond all doubt. In volume xvi Dr. Herman Weber brought before the society an example of embolism of the renal arteries, connected with mitral disease of the heart. See also cases reported in volumes x, xi, and xiii, of renal complications after acute rheumatism.

In all severe cases of acute rheumatic fever, with cardiac complication, the young practitioner must, therefore, be on his guard against the possible detachment of the fibrinous deposit from the valves of the heart, and their transportation to remote organs, the brain, lungs, liver, or kidneys, but he must also be prepared for the possibility of an exudation of fibrine within the inter-tubular vessels of the latter organs, and a subsequent disorganizing process which is equally fatal to life.

§ 7. GOUTY NEPHRITIS.*—It is very doubtful, whether apart from the irritation of gravel or of a calculus in the kidney, conditions commonly coincident with gout, the kidneys are, in gouty patients, susceptible of that kind of subsidiary disturbance, which in the so-called retrocedent form takes place in such organs as the brain, lungs, stomach, heart, intestines, or bladder.

Symptoms of urinary disturbance are doubtless among the

* This condition of kidney whether from gravel, calculus, or any other cause must be cautiously distinguished from what has been called the gouty or atrophic kidney, which is treated of hereafter.

most prominent and most frequent in the gouty patient; but these refer chiefly to the qualities and quantities of the urine. Hæmaturia is a not uncommon symptom in strongly-marked gouty habits; but this appearance of blood in the urine is so constantly associated with the symptoms typical of gravel or calculus, the *débris* of uric acid, either as sand or gravel being always found in the sediment of blood-corpuscles, that there is great reason to believe that the hæmorrhage arises from the irritation of these particles, and not from any diffuse engorgement of the kidneys, which would occur if they were the seat of a special gouty inflammation. Sydenham, in his treatise on gout, referring to the symptoms which most endanger life, and which suddenly appear on the retrocession of the gout, threatening death to the patient, in the phraseology of his time, speaks of the metastasis of the peccant matter to the stomach, intestines, heart, lungs, and brain; but with respect to the kidneys he merely states that all gouty subjects, when they have struggled a long time with the disease, are liable to stone in the kidneys, and to be afflicted with nephritic pains; and, speaking of himself, who had suffered long and severely from bloody urine, arising from a calculus impacted in the kidney, he describes with his usual clearness the most typical symptoms of the nephritic paroxysms caused by the irritation of a calculus.

Scudamore in his work* on gout and rheumatism, makes no further allusion to renal disorder in retrocedent gout, than saying he visited an elderly gouty gentleman under an inflammation of the kidneys, and he merely adds that bleeding from the arm to a free extent was one of the means of treatment attended with the best success. No symptom is mentioned that enables the reader to judge from what the inference was derived, that he suffered from inflammation of the kidneys. Probably it was hæmaturia, and as neither vomiting, nor any symptom usually characteristic of nephritis is recorded, the inference must be that hæmaturia arose from either gravel or calculus.

It is very clear that the term nephritis in relation to gout

* Scudamore 'On Gout and Rheumatism.'

was very vaguely employed by former writers. So high an authority as Scudamore has fallen into this laxity. A case mentioned by him* as one of nephritis, is clearly, from his description, one of nephralgia from gravel. It occurred to a medical gentlemen, fifty years of age, of full habit, and sanguine temperament; he never had gout, but his urine constantly deposited the uric acid sediment. One morning he was suddenly seized with an acute pain in the site of the left kidney. In half an hour vomiting took place, the pain passed downward, in the direction of the ureter, to the left testis, which was strongly retracted. Efforts to pass urine were constant and agonizing, and voided only in teaspoonsful. He was free from fever, and the pulse was not affected. He was bled to ʒxvj. A warm bath almost instantly obtained relief. He had reason to suspect that some gravel passed while in the bath.

These symptoms are clearly not those of nephritis. Their duration, a few hours; the absence of fever, or any acceleration of the pulse tells this sufficiently. The intense pain, the vomiting, the temporarily retained, not suppressed urine, doubtless from spasm, sympathetic with the irritation in the ureter, clearly speak of renal colic, and the sum of the symptoms teach us that the attack should have been called nephralgia, and not nephritis.

Dr. Garrod†, however, thinks that the kidneys appear occasionally to be attacked with gout, by which is meant a deposit of urate of soda in the tubular structure of the kidney. When this takes place, he says there is pain in the back, and other symptoms found in nephritic forms of disease. He further adds, "my own observations lead me to think that here gouty inflammation is often set up in the structure of the kidney, accompanied with deposit of urate of soda, not merely in the tubuli uriniferi, but in the fibrous tissue itself." Dr. Garrod's opinion upon this subject is entitled to so much respect, that I feel some hesitation in expressing my doubt, whether these crystalline deposits, in their earliest isolation

---

* Scudamore 'On Gout,' p. 596.   † 'On Gout and Rheumatism,' p. 505.

in the kidney, give rise to any symptoms beyond those which are so frequent in the gouty subject, and arising from excess of uric acid, sand, or gravel in the urine, namely, nephralgia.

§ 8. CALCULOUS NEPHRITIS—*from uric acid accumulations, sand, gravel, or stone.*—Post-mortem examinations of the kidney have clearly demonstrated that uric acid grit, urate of soda, as well as minute concretions of oxalate of lime, are deposited and retained within the tubular structure of the kidneys. It is the retention of these crystalline grains within the tubular structure which may be regarded as the origin of the renal calculus, and moreover, as explaining the cause of such frequent returns of hæmaturia in the early stage of this disease. When these crystalline grains of uric acid are washed out of the tubes, and they often exist in great abundance, they give rise to the presence of what is called red sand in the urine. The sediment in the urine is made up of an abundant deposit of reddish, or orange-red grains, not unlike Cayenne pepper. This state of urine is accompanied by a great amount of lumbar pain, and irritability of the renal passages. When the uric acid is freely excreted in this form, there is no tendency to hæmaturia; but if the uric acid increases beyond a certain proportion, it no longer appears as isolated amorphous crystalline grains of pure uric acid. Urates of soda and ammonia serve to bind together, in larger masses, these excreta, and gravel, as it is commonly called, makes its appearance, either as minute spheroid granules, varying in size from a poppy seed, or in coarser and more irregular masses. Hæmaturia with more or less constitutional disturbance may accompany this deposit. Great irritability of the renal passages, a frequent desire to micturate, with more or less lumbar pain, and occasionally numbness in the course of the cutaneous nerves of the external part of the thigh. If the stomach becomes irritable, with loss of appetite, and retching, the kidneys may be regarded as in a state of inflammation, and calculous nephritis will be the result.

The gouty habit of body which seems specially characterised

by the excess and retention of uric acid in the blood, is more disposed to the formation of these crystalline grains, and it is in this form of constitution that gravel or sand is most prone to appear in the urine; and hence this form of inflammation of the kidney has been designated gouty nephritis.

Urine which contains a maximum proportion of uric acid, will, on cooling, often deposit an abundance of octahedral crystals of oxalate of lime. The presence of these crystals was formerly thought symptomatic of a peculiar diathesis, then called the oxalic, and Dr. Golding Bird, in his work* on urinary deposits, was the first to announce his belief of the connection between the occurrence of these octahedral crystals and the existence of certain definite ailments characterised by great nervous irritability.

The presence of oxalate of lime in the urine will be the subject of future inquiry. In the mean time, it may be stated that calculi formed in the kidneys, are composed either of uric acid, or oxalate of lime; the latter predominates as the nucleus. The oxalate of lime renal calculus gives rise to a greater amount of irritation than uric acid calculi; and hæmaturia is a much more constant and persistent symptom.

A minute concretion of uric acid, or a still smaller granule of oxalate of lime, retained within the tubular structure of the kidneys, may sooner or later develop symptoms of nephritis. These act as irritants by their presence in the tubes; an inflammatory zone surrounds them; the least disturbance excites a little escape of blood, and hæmaturia small in amount, but of frequent recurrence, is the most common symptom of the early stage of renal calculus.

In some rare cases, however, a renal calculus exists, and even escapes from the kidney, and passes the ureter into the bladder without any indication of its presence, till on passing water an obstruction occurs, and after a period of agonizing efforts to overcome the obstruction in the urethra, a jet of urine carries with force the calculus before it. Its shape and composition proclaiming, without doubt, its renal origin.

* 'Urinary Deposits,' p. 120.

In the generality of cases, the slight amount of renal hæmorrhage, with more or less sympathetic indication from neighbouring parts, with a succession of symptoms which will be fully detailed hereafter, are sufficiently diagnostic.

A calculus may be retained in the kidney for years; it may not be suspected during life, or it may establish such alteration and disorganization of the organ, that scarce a vestige of original renal, tubular, or secreting structure may be left.

Acute nephritic symptoms, in the sense of inflammatory action of an acute form, are rarely excited by the presence of a calculus. The changes of structure, which the kidneys undergo, are of a slow and essentially chronic character, and are often painless. Practically, the structural changes produced in the kidneys by the presence of a calculus, will be dependent on the locality or part of the kidney in which it is lodged. The minute crystalline material which forms the nucleus of the calculus, is, for the most part, formed or found originally in the large straight tubes of the pyramids; a zone of inflammation surrounds it. If still within the course of the urinary current, it increases slowly by fresh accessions of the original material, so that layer after layer of uric acid may be added, or crystalline addition of the oxalate of lime may slowly augment its volume; or in the hæmorrhage which the mass excites from the renal tissues, some portion of blood may adhere to the surface, and thus increase of volume sufficient to enlarge the area of inflammation may be excited. A condensation of the parts immediately surrounding the deposits takes place, a capsule is formed, and for the remainder of life the calculus remains innocuous and unknown perhaps. This must certainly have been the case in the individual from whose kidney this description has been taken. I once removed six smooth, hemp-seed calculi of oxalate of lime, from a pouch embedded in the kidney of a man who died from cholera. The pouch had no outlet, nor any communication with the pelvis of the kidney. It was probably one of the calyces occluded. The next condition is that of calculous pyelitis, where the calculus, from its proximity to the surface, has excited inflammation in one or more of the cup-shaped recesses of the kidney, which, extending from one to the other, ultimately implicates the entire pelvis, reaching even to the infundibulum or funnel-shaped commencement of the ureter. The calculus may become lodged in one of the calyces, and there retained, and there keep up a constant irritation, and a muco-purulent fluid is formed from the irritated surface with which the calculus is in contact. This admixture of pus sets in motion a series of molecular and chemical changes, by which the urine becomes alkaline, and the triple phosphate abundantly forms; it is freely deposited on the surface of the calculus, which now increases in volume more rapidly than heretofore.

The volume of the calculus thus slowly increasing by the deposit of triple phosphate is apt to mould itself to the shape of the cavity of the pelvis, and hence these calculi are branched and irregular in their external form. They become impacted in the pelvis of the kidney; their size bars the free flow of urine from the kidney through the pelvis and infundibulum; the urine accumulates behind the obstruction; a dilating force is exercised; the pressure causes the kidney to become distended; a renal tumour forms, perhaps subsides, when temporarily the pressure has opened a channel once more through the infundibulum. The outlet is again blocked, a renal tumour is again detected, and thus, oftentimes through a series of months, even years, the renal structures undergo a slow change. The dilated kidney becomes pouch-like and sacculated, with a branched irregular-shaped calculus immovably imbedded in the pelvis; the cortical and tubular structures of the kidney entirely disappearing, and the sacculated kidney becoming a mere bag, the wall of which consists of nothing but the capsule of the kidney, denser than in its ordinary state, and perhaps on its inner side preserving some traces of the original structures, and particularly of the subdivisions of the calyces. The contents of these sacculated kidneys vary much—in some it will be a urino-muco-purulent fluid—in which case a renal tumour will have pre-existed. In other cases absorption of the fluid parts may have caused contraction of the once dilated kidney, and the sac may contain a creamy or putty-like mass, in which the glistening scales of cholesterine may be readily seen. These pouched or sacculated kidneys constitute the pyelo-nephrosis of some writers.

It is very difficult to trace to its origin the cause of these calculous formations in the kidneys. The gouty constitution doubtless exercises a paramount influence, but equally, perhaps, does hereditary predisposition impart to the organism a tendency to defective elimination which no errors of diet or luxuriousness in mode of life can equal. Among the many cases of calculous disease of the kidneys which have come under my observation I think nearly one half had a family history in which either gout or renal calculus had been present in one parent or the other, or in some collateral branch derived from the grand-parents. In many of these the habits of life in respect to diet and exercise were simple and innocuous; the general health good and unimpaired; and a complete exemption from those dyspeptic troubles which are the heritage of the man who acquires gout through the gratifications of the palate associated with slothful inactivity of body.

Next, however, to hereditary disposition, in its influence in causing renal calculus, may be reckoned indolent and luxurious habits and modes of life; bodily inactivity has, perhaps, a larger share than the most luxurious diet. It may be said that whatever tends to the development of gout may have a possible tendency to the generation of renal calculus.* Gout is essentially a disease of defective elimination—a detention in the organism of excrementitious material, and a deposit thereof in certain textures, tendinous, cartilaginous, and periosteal. In renal calculus the same kind of excrementitious material is deposited in the kidney; the diathesis of the individual is also strongly assimilated to and associated with gout. The patient with a renal calculus may never have had gout, yet whatever disorders he may have suffered they are of the gouty type. Certain skin eruptions are common to both—psoriasis and

* Sydenham Works, vol. ii, p. 155.

lepra; some forms of eczema are common to the uric acid diathesis, whether in the form of gout or renal calculus.

The remote cause of the defective elimination of uric acid, whether in gout or renal calculus, has been sought for either in an excess of supply of the nitrogenous elements of nutrition, an excess beyond what the assimilative organs can appropriate, and beyond what the organs can eliminate, or, what is more probable, in some defective process of oxidation, by which the uric acid formed in excess should be converted into urea and oxalic acid, and this failing it accumulates in quantities beyond which the blood cannot hold it in solution, and in the solid form is either deposited in the tendinous or cartilaginous textures, as in gouty inflammation, or in the kidneys as sand, gravel, or stone.

Calculous nephritis, then, although frequently occurring to patients who have never suffered gout, is nevertheless intimately connected, as has been already stated, with the gouty constitution.

What is called gouty nephritis, therefore, is a nephritic paroxysm with bloody urine, occurring in the intervals of the attacks of gout and dependent on stone or gravel in the kidneys. The symptoms simulate inflammation as they are ushered in by hæmaturia, and there may be some degree of constitutional disturbance which is, however, dependent rather on the pain and renal colic than on any febrile condition symptomatic of inflammation.

The subject of gouty and calculous nephritis will be further considered under the section of diagnosis and symptoms.

§ 9. NEPHRITIS, or *Pyelo-Nephritis, from stricture and retention of urine.*—Retention of urine from whatever cause, whether arising from disease of the urethra, prostate, or bladder, is recognised as a cause competent to produce nephritis, or rather that condition of the renal organs which has been designated pyelo-nephritis.

Mr. Stanley has recorded a case of the sudden suppression of a gonorrhœal discharge by injection, which was followed by retention of urine, hæmaturia, and death, intense nephritis being caused by the retention of urine. The form of renal disorder usually excited by retention of urine, either as a consequence of stricture or prostatic disease, is, however, of a more chronic character. The retained urine dilates and irritates the ureters, infundibulum, and calyces of the kidneys, a chronic form of inflammatory action is established, at first analogous to a catarrhal state of other mucous surfaces, with more or less mucous fluid mixed with the urine; thickening of the mucous membrane slowly proceeds; the mucus passes into pus, and the urine becomes permanently purulent. The thickening of the mucous membrane of the ureters is frequently irregular, and occurs most at the termination in the bladder, the diameter of the canal becomes narrowed, and the fluids passing from the kidney are partially retained, a dilatory force is slowly exercised, and a pouch-like condition of the ureters, with sacculated state of the pelvis and calyces is the ultimate effect of complete as well as partial, or intermitting attacks of retention of urine.

Examples of these cases are to be seen in all hospital museums. They are frequently classed under the term pyelo-nephritis.

§ 10. NEPHRITIS, *Tubercular or Scrofulous.—Definition.—* A suppurative disease of the kidneys from deposit of tubercle within the substance of the organ, and usually, during life, proceeding without evidence of tubercular disease elsewhere; urine uninterruptedly purulent.

*Tubercular pyelitis.—Definition.—*A suppurative disease of the pelvis of the kidney, the ureters and urinary passages, from tubercular or scrofulous deposit in the submucous layer of those parts, occasionally, but not necessarily, extending to the kidneys, and usually with manifestation of tubercular or scrofulous disease elsewhere; urine mucoid, purulent, and ropy.

Pathology.—It may be necessary to premise that in the following account of tubercular disease of the kidney, tubercular matter and scrofulous matter are considered identical. The elementary and chemical composition, as well as the changes, or the metamorphosis, of each being in all respects similar.

In Rokitansky's scale of the frequency of tubercle in the various organs and textures of the body, tubercle of the kidney occupies the ninth place; the kidneys therefore are not frequently the seat of tubercular deposit.

Tubercular disease of the kidneys is not always accompanied by tubercular activity in other organs. Pulmonary phthisis and renal phthisis do not often run their course simultaneously. In very extreme cases of tuberculosis, however, tubercle may be diffused through all the chief organs of the chest and abdomen; the lungs disorganized by tubercular deposit and ulcerations; the intestines and mesentery infiltrated with tubercle; one or both kidneys the seat of tubercular softening and purulent change; the ureters the seat of the same deposit, and extending to the bladder, infecting the prostate and canal of the urethra, implicating both the membranous and spongy portions of the urethra, and thus carrying, by the agency of secondary deposits, the tubercular disease, to the utmost limits of the urinary apparatus.*

Such extreme diffusion of tubercular disease, with a corresponding degree of destructive activity simultaneously proceeding, is an exception to what is ordinarily observed in tubercle of the kidney. Most commonly, tubercular nephritis and pyelitis run a rapid and unchecked course without evidence of tubercular disease in other organs. The object of this section is to describe the symptoms and morbid anatomy of that form of renal disorder which is caused by tubercular deposit in the kidney, laying aside for the present all reference to tubercular disease in other organs, or the possible complications which may coexist with tubercle of the kidney.

When the renal symptoms have from the first been the most prominent, the tubercular disorder is usually limited to

* See 'Pathol. Trans.,' vol. xi, p. 137, 138.

the kidneys, having its origin in them, and extending by secondary infiltration to the pelvis and outlets. In such cases, after death, no indication of tubercle in other organs may be seen. The disease may be thus limited to the renal apparatus. On the other hand, tubercle may occur in both lungs and kidney, but in different degrees of activity. Fatal cases of acute pulmonary phthisis have been observed in which tubercle has been found deposited in both kidneys, and which had advanced to a certain stage of softening, even to the formation of small abscesses, but the patient had died of the pulmonary phthisis before the ulcerative process in the kidney had established an outlet, so that no indication of such complication existed during life.

The cortical part of the kidney is the primary seat of tubercle; it is in the form of yellow tubercle, and hence, as occurs with this variety, it runs a rapid and acute course. Numerous yellow points are seen surrounded by an inflammatory zone, in the centre of which is a minute drop of tubercle pus. In the same kidney may be seen tubercular spots in varying degrees of progress :—1st. The tubercular puncta with their inflammatory halo or zone. 2nd. Small cup-shaped cavities, out of which a minute quantity of yellow creamy pus may be emptied. And 3rd. The fully developed tubercular abscess, scooping out by its eroding action into irregular hollows the base and substance of the cones, and emptying the purulent formation into the infundibulum mixed with urine, secreted by those portions of the organ not yet destroyed by the ulcerative process. The disease may be confined to one or extend to both kidneys. It runs a rapid and for the most part unchecked course. The tubercular disease may be limited to the renal structures, that is, it may not pass beyond the cortical and medullary substance of the organ, but in the great majority of cases, by secondary deposits, the pelvis and ureters become involved, and it is to the implication of these latter, and the thickening of their walls and narrowing of the diameter of the canal, that the foundation is laid for that obstruction to the free discharge of the pus and urine which eventually terminates in the formation of a renal tumour or renal abscess.

Renal tubercle advances to the stage of softening and purulent destruction of surrounding texture by a process strictly analogous to what has been observed in yellow pulmonary tubercle. The commencement of its destructive activity is marked by a zone or area of inflammatory redness surrounding the yellow tubercular spot. It softens from the centre, becomes cheesy, then creamy, and as pus cells become developed at the expense of the tissue around a still wider area of inflammatory engorgement is formed, in which fresh tubercular matter is actively deposited. The area of engorgement thus goes on increasing, and with it an extending arc of tubercular infiltration; destruction of texture proceeds till an outlet is formed into one of the calyces, and the softened tubercle and accompanying pus are voided through the ordinary urinary passages. All parts of the urinary apparatus may thus become infected with tubercular infiltration—ureters, bladder, prostate, in the male,—ureters, bladder, and ovary, in the female; but these are secondary results, and arise from the extension of the primary tubercle deposit in the kidneys to the urinary passages and contiguous parts. The extension of the disease to the bladder in some measure alters the character of the urine and tends sometimes materially to obscure the diagnosis. So long as the tubercular ulceration is limited to the kidney the urine is simply purulent; let at rest it separates quickly into two portions, a well-defined sediment of pus cells and a clear upper fluid of urine, faintly acid and slightly albuminous, derived from the liquor puris. But when the bladder becomes implicated in the tubercular mischief the urinary secretion is voided ropy, or quickly becomes so, rapidly decomposes and develops, often immediately after passing crystals of the double phosphate of ammonia and magnesia. These conditions may beget the belief, particularly in the absence of a renal tumour, that the bladder is the seat of the mischief, and that a calculus may be the source of irritation. Hence it is apparent that a trustworthy account of the antecedent symptoms, and the order of their sequence, is essential to a correct diagnosis. An examination of the bladder would disprove the existence of calculus, but it would not exclude other possible sources of bladder disorder.

In rare cases the deposit of scrofulous or tubercular matter has been found limited to the ureters and bladder, the kidneys being free from tubercular taint. Such conditions usually occur in the course of scrofulous disease of other parts, particularly of the joints in young people. The urinary disorder is not primary; it manifests itself only during the progress of the disease of the joints. Incontinence of urine or distressing frequency of micturition, irritation or pain at the neck of the bladder or urethra, the urine being cloudy, purulent, and ropy, are the usual symptoms of some scrofulous affection of the urinary apparatus, and may be produced by tubercular or scrofulous deposit limited to the outlets of the kidney. The exact seat of this deposit has been determined by Dr. Handfield Jones to be the sub-mucous tissue, or beneath the epithelial lining of the ureters and pelvis, that is, beneath the layer of cells and the basement membrane on which it rests.*

On examining a vertical section of the ureter, in the case of a boy, aged eleven years, who died from scrofulous disease of the hip-joint, with tubercle in the cerebellum, a small one on the liver, none in the lungs nor in the kidneys, but in the left ureter a small deposit of scrofulous matter was observed, which was distinctly seen to be deposited subjacent to the line of epithelium.

The diagnosis of these cases is often difficult; the extension of the deposit in the direction of the bladder causes the urine to become purulent and ropy; the pain and suffering of the patient is often expressed by a general irritation of the urethra and neck of the bladder. The age of the patient may prevent any accurate account of the special subjective symptoms, and these are frequently strongly suggestive of calculus in the bladder; an examination is made and none found. With

* "On the Minute Anatomy of Scrofulous Deposit in the Ureters." Dr. H. Jones. 'Trans. of the Pathological Soc.,' vol. i, p. 283.

evidence of scrofulous disease in other parts, the absence of proof of vesical calculus, the persistence of the urine purulent and ropy, with abundant crystals of the double and triple phosphate as microscopic objects in the urine, experience would justify a diagnosis of scrofulous pyelitis or of some tubercular disease of the urinary organs beyond the limits of the kidney.

Morbid anatomy.—Tubercular and scrofulous disease of the kidneys consists in the deposit of a material which, after passing through a certain state of changes, excites in and around its position inflammatory action, with the formation of pus and destruction of tissue; the appearances after death presented by the kidneys exemplify the range of the ulcerative process, its duration, the free or obstructed flow of the purulent urine, the extent to which tubercle may have infiltrated beyond the limits of the kidneys, and lastly, the possible obsolescence of the tubercle pus, and its conversion into a putty-like mass, consisting of a dense granular matter, with the presence of cholesterine in quantity sufficient to impart to the material a glistening spermaceti-like character.

For conciseness' sake it may be convenient to speak of four varieties or appearances of the kidneys which tubercular disease may display.

Two forms may be noticed occurring in cases in which the tubercular disease of the kidney is directly fatal by asthenia. Two other forms are met with in cases where death has been occasioned by other causes, and where no symptom of renal disorder has immediately preceded death.

Of the two forms in which the ulcerative process is the direct cause of death, the difference consists essentially in the free passage of the pus through the ureters. When this canal is free from tubercular complication the kidneys do not become sacculated, and there is no renal tumor. In this form both kidneys are usually the seat of disease, one, perhaps, somewhat in advance of the other, but tubercular cavities or abscesses occur in both. The kidneys when removed have an elastic or boggy feel, and on a section being made a number of uneven, irregular cavities are seen freely opening into the pelvis of the kidney.

These abscesses vary in size and position, some approach the very limits of the cortical substance, others appear at the base of the pyramids, and others at the apex of a cone, breaking up and destroying the secreting substance. Nevertheless, there is abundant evidence of large portions of kidney tubular structure remaining unaffected, and capable of carrying on some portion of excreting power.

In the cortical part, and also at the base of the pyramids, minute spots of yellow colour, surrounded by a zone of redness, may be seen, in the centre of which a small quantity of creamy fluid may be emptied, leaving a central cavity as small as the smallest mustard seed; these may be considered as examples of the earliest stage of tubercular abscess. In such kidneys a considerable portion of tubular and secreting structure remains unimplicated. The urinary tubules when examined by the microscope do not present any condition of the epithelial gland cell characteristic in any way of diseased action. In the cases under consideration the tubercular disease is confined to the limits of the kidneys, and has not extended into any of the passages or outlets beyond. The apex of those cones not yet implicated do not differ in appearance from what may be observed in healthy kidneys.

The lining membrane of the calyces and of the infundibulum, and throughout the upper part of the ureters, is generally very highly marked with arborescent vascularity, but no evidence of tubercular infiltration or ulceration is noticed. In the second form of tubercular kidney the great mass of the organ has been destroyed by the ulcerative process, and the deposit has extended beyond the limits of the kidney, has invaded the pelvis, involved the mucous surface, and carried even to the ureters the tubercular deposit. The walls of the ureters become thickened, the diameter of the canal narrowed, and slowly the flow of purulent urine is impeded. A distending force is thus brought to bear on the interior cavity of the organ, proportioned to the impediment to the flow of the constantly accumulating purulent fluid, and becoming greater as the obstruction increases, exercising a twofold influence, one accelerating the ulcerative process by constant pressure and contact of the pus with the substance of the organ, the other dilating with a gradually increasing force the yielding walls of the slowly destroyed kidney. When the ureter is completely blocked, which occasionally happens, the kidney becomes converted into a sac, or sacculated, that is, divided into compartments or cavities, caused by the resistance of the fibro-cellular stroma to the destructive action of the ulcerative process.

The walls of the smaller cavities are usually ragged with membranous shreds. The pelvis, infundibulum, and ureter are also covered with false membrane, the latter has its walls much thickened, and becomes cord-like, and its canal so limited, as to permit but a small quantity of fluid to pass through it. This seems brought about partly by a diffuse inflammatory process, as well as by the deposit of tubercular or scrofulous matter beneath the epithelial layer and the basement membrane on which it rests.

Complete obliteration of the canal is not so common as in calculous pyelitis.

The size of these encysted kidneys varies. In most cases, the tumour can be easily felt through the abdominal walls, and with the greater facility from the emaciation of the patient and the attenuation of the abdomen.

It may extend on the right side from the inferior margin of the liver downwards to the iliac fossa; on the left, from the lower edge of the spleen to the like position inferiorly. Pressure made on it during life produces a sense of sickness, with a pungent darting pain to the pubis or neck of the bladder.

The disposition to ulcerate outwards, and to establish a communication either with the intestines or with the external walls posteriorly, is not so frequent with this form of renal abscess as that produced by calculous disease.

The other two forms of tubercular or scrofulous kidney are

met with in cases where death has been caused by other disease than that of the kidney.

The first occurs in cases where pulmonary tubercle has advanced and proved fatal by the usual destructive process in the lungs, and where the kidneys have also been the seat of tubercular deposit, but which has not advanced beyond the stage of softening, and when, although pus has been formed, no communication with the renal outlets has been established, and consequently no system of urinary disorder has existed during life. Such cases are extremely rare; the writer has, however, met with two such examples—one in the practice of Mr. Kingsford, of Sunbury, the other among the patients of the hospital to which he is attached. The kidneys in both these cases presented similar appearances. On the tunic being removed, the cortical surface exhibited numerous yellow spots of tubercular deposit, and on a section being made, several small collections of pus, having no communication with each other, nor with the pelvis of the kidney, were observed. The creamy yellow pus being removed, a small cup-shaped cavity was left, the walls of which were formed by a firm red margin or wall of condensed structure. No indication of such a state of kidneys exists during life. In both these patients extensive tubercular ulceration of lung substance existed, and was the cause of death.

The remaining form of scrofulous kidney is equally rare; it presents the unusual condition of a collection of a soft, putty-like substance contained within the pelvis of the kidney, the ureter being closed, or the material is encysted, and confined within one of the calyces.

It appears to be scrofulous or tubercular pus, the more fluid portion of which having been removed by absorption, the cells and their fatty nuclei undergo some molecular change, leaving a fine granular mass through which is diffused a large proportion of cholesterine. This mass contained within the kidneys, to the unaided eye, glistens, as if sprinkled with spermaceti, and the microscope discovers the characteristic rhomboidal plates of this form of crystalline fat.

It affords an example of the change which takes place in tubercle or scrofulous pus by the resorption of the serum and the fatty metamorphosis of the cell elements.

This process seems to imply the disappearance or extinction of what Rokitansky calls the tubercular dyscrasia. The disposition to deposit tubercle or scrofulous material has ceased on the return of higher and more conservative nutritive processes in the organism, and with the cessation of the pus-forming (pyo-genetic) causes, a retrograde process commences, the fluid parts of the pus are removed by absorption, and a material remains innocuous and inert, composed of a fine granular material through which the crystalline rhomboid plates of cholesterine are diffused. These examples have been met with in individuals who have died from other disorders than those of the kidney, and probably a long time after the material was formed.

An example of this form of disorder is mentioned in the fifth volume of the 'Transactions of the Pathological Society,' p. 178. It was a case which occurred in St. Thomas's Hospital under Dr. Peacock. The patient suffered from pulmonary and cardiac symptoms, but no sign of tubercle was met with in other organs. The right kidney was small and lobulated; the cortical and medullary portions were completely destroyed, and the space occupied by cavities varying in size; they were filled with an opaque white substance of uniform character, soft, but not fluid, and resembling moist plaster of Paris. The ureter was obliterated about an inch from the hilum, and from this point to the bladder it was reduced in calibre.

The putty-like material under the microscope presented a large number of imperfect broken-down cells having much the appearance of pus mixed with a quantity of fine granular matter, small oil-globules, and occasional plates of cholesterine. Dr. Bristowe, who reported on the case, states that the above changes were like those which take place in tubercular disease of the kidneys, but he inferred, from the absence of tubercle elsewhere, that it was simply consecutive pus, the result of bygone suppurative inflammation of the secreting structure of the kidney. I would venture to differ from this conclusion from the fact that suppurative inflammation of the kidney, apart from calculous or tubercular origin, is very rare without giving rise to symptoms specially characteristic of renal disorder during life, while the absence of tubercle in other organs would lend strength to the supposition that this material had its origin in tubercular pus, its inertness and subsequent metamorphosis depending rather on the extinction of the tubercular crasis, the tubercular deposits never having passed beyond the limits of the kidney.

§ 11. NEPHRITIS—*From cancerous deposit; cancer of the kidney.*—The term cancerous nephritis is not perhaps pathologically correct. It is doubtful to me whether the deposit of cancer in the kidneys is capable of exciting diffuse inflammation of the kidneys; at most a zone of inflammatory action may be established around the spot in which the malignant deposit is formed. There are certainly no traces of inflammatory action in the ordinary sense of that term. The cancerous deposit in these organs, as elsewhere, undergoes certain ulterior changes according to its specific character. The chief form of renal cancer is the encephaloid; the fibrous or hard schirrous quite exceptional. The cancerous deposit passes through the usual stages of softening and suppuration, in the progress of which this degenerating deposit may be mixed with the urine in the bladder, or a renal tumour may form and the abscess find its way outwardly in one of several directions. It may find its way into the colon, and the contents be discharged per anum; it has ulcerated its way into the duodenum, into the cavity of the peritoneum. Under the head of diagnosis and symptoms of cancer of the kidneys this subject will be further considered.

§ 12. NEPHRITIS—*endemic; endemic hæmaturia.*—Certain climates are now known to be favorable to the development of nephritis. Hæmaturia, it is well known, prevails as an endemic in the Mauritius, certain parts of the Cape Colony, and in some regions of South America, as well as in Egypt. This hæmorrhage has, however, now been traced to parasitic agency of an entozoon, which has been named *Bilharzia hæmatobia*, after its discoverer Dr. Bilharz.

See an interesting paper in the 'Med.-Chir. Trans.' by Dr. John Harley (vol. xlvii, p. 55 *et seq.*). In that paper Dr. John Harley gives a minute description of the ova, composed of the immature embryo contained in an egg case, as well as the mature embryo, with drawings illustrating the progress of development of this entozoon, which has by him, as well as by Dr. Cobbold, been identified with the hematoid *Bilharzia hæmatobia*. It is the cause of the endemic hæmaturia and gravel complaint or lithiasis of Egypt. So frequently does the disease exist in this latter country that Griesinger found it in 117 out of 363 post-mortem examinations. It chiefly inhabits the small vessels of the mucous membrane of the urinary organs, as well as the kidneys themselves.

A more recent communication by the same author on this subject was made to the Medical and Chirurgical Society during this session of 1869. The parasite, from further information conveyed to Dr. John Harley, gains access to the body chiefly in the act of bathing in the rivers; and it is an interesting fact, that Europeans resident in the interior of the Cape are free, while, if they change their residence to the coast, soon become infected. The ravages of the parasite are now shown to extend beyond the mucous membrane of the urethra and bladder, and to invade the kidney.*

This form of renal disorder, now beyond all doubt traced to the presence of a hematoid worm, the *Bilharzia hæmatobia*, was at first referred to the ravages of the parasite upon the urethral and vesical mucous membranes only. But Dr. John Harley, in the paper already quoted, has already received information that in one case at least the parasite had extended to the kidney, and renal calculi were occasionally passed. And

---

* 'Proceedings of the Medical and Chirurgical Society,' xxix, p. 214.

he has also detected in these calculi, which are composed of oxalate of lime, considerable numbers of ova which without doubt afforded a nucleus for the crystalline deposit; so that there can be no hesitation in tracing to this parasite renal as well as calculous disease.

§ 13. NEPHRITIS *in connection with pregnancy.*—Hæmaturia may occur in women as a form of vicarious menstruation, but as this is but a temporary symptom, and is not succeeded by any of the consequences of inflammation, it should not be regarded as a form of nephritis. Indeed this may be said of hæmaturia arising from some other causes, particularly the action of turpentine or copaiba, as these agents induce but a state of hyperæmia or blood stasis which the hæmaturia relieves.

I have elsewhere* treated of albuminous urine in connection with pregnancy. But the present work in its scheme would be incomplete if reference were not made to the occurrence of a congested state of the kidneys during the progress of uterogestation. The pressure of the gravid uterus on the abdominal vessels, particularly on the iliac veins, is sufficient to explain the swollen feet and œdematous ankles so frequently observed in the latter months of pregnancy. But it occasionally happens that the urine becomes albuminous, and that the anasarca extends beyond the limits usually assigned to the pressure on the great trunk veins which receive the blood from the lower extremities. The ascent of the gravid uterus beyond the limits of the pelvis may extend its pressure to the efferent vessels of the kidney, and the circulation through these organs be sufficiently impeded to produce albuminous urine. Even hæmaturia may add to the gravity of the symptoms, and a diffused anasarca may excite apprehensions of that formidable complication of Bright's disease with pregnancy. In one class of cases where the urine has been albuminous the patient is safely delivered, the dropsy disappears, the urine becomes free and natural, and convalescence steadily proceeds,

---

* 'On Dropsy,' 3rd edition, cap. xvi, p. 230.

leaving on us the conviction that these alarming conditions originated in mechanical pressure and not in any special constitutional form of renal disease. But there are other cases, not infrequently met with in obstetric practice, in which the renal disorder is not thus secondarily induced, but, on the contrary, the period of delivery is the period of danger; convulsions occur either during or after delivery, and but too frequently with fatal results.

I am inclined to view these cases as originating not in pregnancy only, but in an antecedent renal condition which required but the disturbance in the equilibrium of the renal circulation which the pressure of the gravid uterus excites to give it activity. Some authorities think that pregnancy is a fertile source of renal disease. My own view is that pregnancy is only an exciting cause in those women already constitutionally predisposed to renal disease, and that pregnancy in a perfectly healthy young woman is not more likely to induce morbus Brightii than it is to bring on disease in any other of the abdominal organs.

§ 14. PERI-NEPHRITIS.—By this is meant inflammation of the tissues surrounding the kidneys. It is not a primary but a secondary affection. It has been known to follow an intense acute nephritis by the extension of the inflammation of the cortical portion of the kidney to the capsule, and thence to surrounding structures. The infiltration of the urine, or of pus from a sacculated kidney into the contiguous structures, contusions on the loins, may provoke inflammation in and about the kidneys. It is a condition very difficult of diagnosis; when it follows pyelitis it will be preceded by symptoms of renal abscess—the ureter has become occluded, a tumour forms, and as the contents are obstructed in their passage to the bladder the purulent fluid burrows through and among the surrounding textures, and a diffused inflammation is ultimately concentrated at a point in the integuments through which, by ulceration or by the aid of a lancet, the pent-up contents of the renal abscess eventually escape. In other cases an inflamed state of the kidney diffuses itself outwards, involves the capsule, and extends to the adipose substance surrounding the organ. Such cases, except as the result of injury to the kidneys by violence, are rare.

CHAPTER III.

NEPHRITIS—SYMPTOMS, DIAGNOSIS, AND TREATMENT.

Inflammation of the Kidneys, with Bloody, Albuminous, or Purulent Urine.

VARIETIES:—1. Idiopathic (doubtful). 2. From external injury. 3. From poisons taken internally. 4. The sequel of scarlet fever. 5. The sequel of other fevers and diseases. 6. From the effects of cold and wet—acute morbus Brightii—inflammatory albuminuria. 7. Rheumatic nephritis. 8. Gouty nephritis. 9. Gouty nephralgia. 10. Lithiasis, gravel. 11. From calculus or stone in the kidney. 12. From tubercle or scrofula. 13. From cancer. 14. From endemic causes, parasitic. 15. Peri-nephritis.

In passing in review the several causes which are competent to produce inflammation of the kidneys, or conditions somewhat analogous thereto, it has been almost impossible to avoid mixing up with the description of a cause some of the symptoms which a given agent may excite. In the following statement of symptoms some unnecessary repetition therefore may be apparent. But as the object of the present work is to assist the inexperienced in the study of obscure diseases, an arrangement of the subject matter of the work upon a plan suitable, not only for ready reference, but equally capable of impressing on the student the force of the differential method of diagnosis, has been chosen, which cannot, without risk of perspicuity, be free from the apparent weakness of repeating what has been before stated.

NEPHRITIS.—*Symptoms.*—Before describing in the order

in which the causes of nephritis have been placed the symptoms peculiar to each, and on which a correct diagnosis must depend, it will be well to state generally the symptoms which are common to every form of nephritis, and then in detail will follow those symptoms which are peculiar to any particular form. This will constitute the basis of the differential diagnosis on which rests the correct estimate of the disease; and the especial object and aim of clinical teaching. In all cases of nephritis, from whatever cause, there will be present at the outset more or less of constitutional disturbance in the form of chills or rigors, followed by flushes of heat and febrile excitement, inappetency, thirst, and general prostration of strength and muscular activity. The stomach is irritable from sympathy with the kidneys, and there may be nausea, even vomiting. The pulse is quick, sharp, and small. In cases of constitutional origin the temperature of the body rises with the febrile disturbance; while in nephritis from external violence or injury the temperature falls below the usual standard, and the extremities, both hands and feet, are clammy and cold. There is more or less pain across the region of the kidneys, which may extend downwards to the crest of the ilium, or to the pubis, or the perineum, according to the nature of the cause. There is a distressing anxiety to pass water, and frequent ineffectual efforts to expel it. The urine is in all cases at first scanty, in some even suppressed, and it is bloody and albuminous, becoming purulent as the disease progresses. Such is an outline of the symptoms which may be accepted as typical of nephritis from any cause. But inflammation of the kidneys is followed by structural changes which vary very materially, though the symptoms do not differ to so great an extent. Yet these differences, trifling as they may at first appear, are sufficient to guide the observer to a correct opinion as to the ultimate nature of the structural changes in the kidneys.

Nephritis from injury or external violence may terminate in complete recovery; or an inflammatory disorganizing process may be established with results not dissimilar to what are observed in the parenchymatous inflammation of the lungs in pneumonia.

In the nephritis—after scarlet fever, or excited by cold and wet, and accompanied by dropsy (acute morbus Brightii), the pathological result, as already shown, is an infiltration into the tubular and inter-tubular structures of the kidney of a granular and fatty product of cell disintegration, with defective excretory power of the chief elements of the urine. In nephritis excited by the presence of a calculus in the kidney, a zone of inflammatory action around the deposit, often first manifested by trifling or frequently recurring hæmaturia, is followed by the formation of a small defined abscess around the calculus, with the presence of pus in the urine; or the calculus may become surrounded or enclosed in a well-defined wall of condensed tissue.* And, lastly, tubercular or scrofulous nephritis has its own specific pathological change, differing in many particulars from any of the former. Each form of nephritis may terminate, therefore, in a different result. To facilitate the diagnosis of these varieties of renal disease is the object of the present work.

§ 1. NEPHRITIS.—*Idiopathic.*—A doubt has already been expressed whether the kidneys are susceptible of inflammation apart from any of the exciting causes previously enumerated. In attempting to establish a correct diagnosis of such cases the typical symptoms of acute albuminuria, of calculous nephritis, and tubercular nephritis, must be carefully kept in view. I have seen but two cases which, judging by the result, would be justly designated under this term. One is mentioned in my work 'On Dropsy,' and the second has occurred since that edition was published. Both cases were young women. The last, twenty-two years of age, whose general health had been good to within the fortnight previous to my seeing her, had at that

* See two preparations of calculi imbedded in the substance of the kidney in the Museum of the Westminster Hospital. In other forms of calculous disease of the kidney the entire structure may be scooped out, and the kidney converted into a pouched or hollow cavity. Examples of these sacculated kidneys may be seen in every hospital museum.

time after getting wet at a country fair suffered some rigors, with retching and vomiting, dull heavy pain across the loins, rapid pulse, elevated temperature, and scanty all but suppressed urine which was bloody. A constant desire to pass water accompanied by this tearing aching pain at the pubis were the chief symptoms. When I first saw her there was great frequency of pulse and a desire to pass water every quarter or half hour. The stomach was less irritable than at first. The urine was milky when passed; at rest it separated into a clear upper portion, and a yellowish sediment on the surface of which was a red or pinkish layer. This consisted of scattered blood discs, the sediment of pus cells, and the clear upper fluid was acid, and slightly albuminous. Some pus casts were also seen. There was nothing in her family history connected with phthisis or struma. There were no sympathetic pains beyond the aching at the neck of the bladder, no pain extending down the ureters, some pain across the loins, and a doubtful increase of pain on pressure over the region of the kidneys. I believed the case to be one of tubercle of the kidneys. The hæmaturia and general febrile condition with vomiting so rapidly followed by purulent urine is a marked symptom in tubercular nephritis. In acute morbus Brightii the hæmaturia is followed by fibrinous and epithelial casts, rarely a rapid formation of pus, and there is usually some anasarca or dropsy. In calculous nephritis, the sympathetic sensations in alliance with the kidney are well marked, pain down the ureters to the neck of the bladder, aching on the crest of the ilium, anterior or posterior, and some discoloured sensibility in the external cutaneous nerve. Moreover the stage of hæmaturia, not usually accompanied by suppression of urine, is slowly followed by mucous urine, a slightly flocculent cloud of mucus, slowly exchanged for pus cells covering periods of weeks or months.

The diagnostic symptom of tubercle in the kidney is hæmaturia, scanty urine, rapidly followed (a few days) by purulent urine which increases, and rarely intermits in quantity. The case of nephritis I have just described had all these distinguishing symptoms, and I gave an opinion in accordance therewith. The patient lived in the village in which I reside in the summer, and I saw her weekly for three or four months.

Opiate anodynes were first prescribed—the paregoric tincture. Occasionally, to relieve the distressing frequency of micturition, a grain of the acetate of morphia at bedtime. She took also cod-liver oil. She did not keep her bed for more than a month. About the sixth week from the commencement of the illness the amount of pus in the urine began sensibly to abate. There was but a trace of albumen. The desire to pass water so frequently had subsided. The sediment consisted of pure pus-cells—no pus casts could be seen. Her general health perceptibly improved. Before the end of the third month the urinary sediment was scarcely greater than in urine containing a little excess of mucus, no albumen present; a few pus-cells with a rather greater proportion of epithelium from the pelvis, ureter, and urethra were the objects visible. Her health was subsequently completely restored; and the urine when last examined was free from all indication of disease. The symptoms of this case differ in nothing from tubercle of the kidney, except in the result. Experience forbids us to hope that tubercular disease of the kidney ever terminates favorably. This case, then, as well as that mentioned in a former work, if not tubercle of the kidney, must be classed as idiopathic, and we must thus admit that the kidneys are susceptible of a parenchymatous inflammation, as the lungs are, in which the products of inflammatory action undergo, as in the pulmonary organs, purulent liquefaction, and that such a disease is not incompatible with complete restoration to health.

§ 2. NEPHRITIS.—*External injury.*—The most prominent symptoms indicative of renal injury by external violence are, a considerable degree of collapse and prostration, feeble, scarcely perceptible pulse, temperature below the natural standard, retching, vomiting, and suppression of urine; if any be passed or drawn off it is scanty and bloody, and in some cases little more than pure blood. The external local evidences of injury may not be very obvious. If blows or kicks have been the cause, bruised spots or ecchymoses on the loins or parts adjacent may be noticed, or pressure over them, or in front, or laterally may produce pain and uneasiness. If the renal mischief has been caused by some violent personal effort, as in wrestling, jumping, or by a fall from a height, there will be no external evidence of injury. The vomiting, the suppressed and bloody urine, will be sufficient evidence of the effect of the injury on the kidneys.

Some years since a policeman of the B division was admitted into the Westminster Hospital under my care suffering from scanty, all but suppressed and bloody urine. He stated he had been ill for two days and treated by the divisional surgeon at his own lodgings. He had been engaged the day before in a street row with some of the Westminster roughs, among whom was a notorious vagabond, remarkable for his strength and predilection for resisting the police. In the fracas the policeman was down, trampled on, and kicked in the loins. He was taken up exhausted and faint, and carried home, where he vomited, and complained of great pain across the loins. Rest and fomentations were enjoined. The suppression of urine and its bloody condition continuing he was removed to the hospital. The chief symptoms on admission were a very feeble weak pulse; a hollow sunken expression of countenance; constant irritability of the stomach and vomiting. The amount of urine passed was not more than two or three ounces and seemed more than half blood. There were no traces of external injury; but great tenderness on the lumbar space as well as by deep pressure anteriorly and laterally. He was cupped over the loins, complete rest enjoined, and hot fomentations applied to the abdomen. Diluents were ordered. The blood gradually disappeared from the urine. In a week blood could not be detected except by microscopic examination. The stomach quickly recovered its power by taking nourishment. Fourteen days after admission and three weeks after the injury the man was convalescent, the remaining symptoms being some degree of debility, and the urine contained for some days an excessive amount of mucus with a few pus-corpuscles and a minute quantity of albumen. Ten weeks after the injury the urine was apparently healthy.

Such are the symptoms usually found in cases of injury to the loins where the kidneys are implicated. The diagnostic symptoms are some degree of collapse, retching, and vomiting, with scanty bloody urine immediate upon some external violence. The microscope reveals little else but blood-corpuscles and fibrinous shreds (coagula). The treatment in all such cases may be summed up in rest, the recumbent position, local depletion, and warm fomentations. Medicinal remedies in these cases of renal injury are of little value. Complications in the course of such injuries may demand other remedial measures. It is very certain that the kidneys may sustain a considerable amount of injury from external violence and yet be reparable.*

While these pages are passing through the press, I have read with much interest some clinical remarks by Mr. Curling on a case of severe rupture of the kidney, with recovery, in the 'British Medical Journal,' for May, 1869. That case affords another example of the extent to which the kidney may be injured and yet compatible with recovery. All the typical symptoms of renal injury were observed, collapse, retching and vomiting, with bloody urine. Opium was needed to allay pain. Cupping was employed. Ice and cold effervescing stimulants allayed the vomiting. The perchloride of iron was eventually given. A purulent and phosphatic urine clearly proved the inflammatory condition of the kidney. Severe as the injury to the kidney in this case must have been, yet by judicious treatment in less than ten weeks the patient was free from all symptoms of the accident and his urine was clear and natural.

Injury to the kidneys caused by some violent physical or muscular effort, wrestling, jumping, or some such similar

* See 'Pathological Transactions,' vol. xi, p. 140, a case of old rupture of the kidney healed, by Mr. Holmes.

strain.—It is very doubtful if any physical or bodily effort of the character just named be capable, apart from some special antecedent condition of the kidney, of producing inflammation of these organs. Cases of hæmaturia immediately or quickly following some such accidental bodily effort are very common, and but too frequently the case is set down as injury to the kidney by such accident or bodily act. It will be found, however, in the great majority of such cases, that either a renal calculus has been displaced in its bed, or that a tubercular or scrofulous deposit has pre-existed in the kidneys, and the bodily act is but the starting-point in the activity of these several disorders. The earliest cognisable symptom in most cases of renal calculus is hæmaturia; and the patient almost invariably attributes the appearance of blood to some accidental bodily act. The subsequent progress of the symptoms sufficiently declares the calculus to have been the cause of the hæmorrhage. It may, therefore, be generally suspected in any case of hæmaturia, in which there are no other constitutional symptoms of disturbance, that the cause is not the bodily wrench or jar *per se*, but that the bodily shock has displaced a substance in the kidney which had remained inactive and unknown till disturbed from its seat.

Even in tubercle of the kidney the first manifestation of renal disturbance is sometimes hæmaturia; and this appearance of blood is often attributed to some accident. The blood, however, in these cases is very early mixed with shreddy membranous masses, and even pus. Hæmaturia, therefore, as a symptom of renal disturbance, after any violent effort, is not necessarily referrible to the direct agency of the violence on the kidneys. If after any such shock or blow retching or vomiting precedes or accompanies the appearance of blood in the urine, and if the urine be scanty, and there be some degree of bodily depression, it may be fairly concluded that mechanical injury to the kidney has been the cause. If, on the other hand, the blood in the urine is unaccompanied by any constitutional symptoms, with only some sense of pain, or weight, or uneasiness about the loins, then the inference would be that a displaced calculus was the cause, and not mechanical injury. A gentleman, of healthy constitution, and of active habits, fifty-four years of age, very fond of field sports, particularly of hunting, who had never suffered a day's illness, had complained of what he conceived was lumbago for some previous days. In the hunting field his horse swerved at a fence, and he felt a kind of wrench at his back. On the same day he was alarmed at passing some blood with his water, and for two or three days following the urine contained blood. He suffered no other symptom, and gradually the urine returned to its natural appearance. The hæmaturia was attributed to the accidental wrench on horseback. When I saw him, two months afterwards, there was no doubt but that the hæmorrhage arose from a displaced calculus, and not from mechanical violence. The hæmaturia continued at intervals of two or three weeks for some months afterwards, but without any constitutional disturbance. In many similar cases the hæmaturia which recurs with greater or less frequency at first, every ten days or a fortnight, is unaccompanied by any of the typical symptoms of renal calculus.*

The treatment of these cases of hæmaturia from some apparent jar or wrench, is similar in all respects to that required for the same symptom after more direct mechanical injury to the loins. Complete rest, warm baths, cupping from the loins, general laxatives, and soothing anodynes, according to the presence of particular symptoms.

Minute attention should be given to all the subjective symptoms or sympathetic sensations which may lead to the suspicion of renal calculus. In the absence of any of the ordinary typical signs, the subsidence of the hæmaturia in a day or two, the urine continuing apparently clear and natural, and then, after an interval of a week or two, the recurrence of hæmaturia may be accepted as all but unequivocal of the presence of a stone in the kidney. The microscope will generally confirm this inference.

* This subject will be fully treated under the head of renal calculus.

§ 3. NEPHRITIS *from Poisons, Turpentine, Cantharides, Copaiba, Nitre, taken internally or externally applied. Lead and Phosphorus act only exceptionally.*—In treating of those causes of nephritis which are comprised under the above head, the susceptibility of some constitutions over others to the influence of these agents was mentioned.

Turpentine applied to the surface of the body produces rarely any other symptom than the increased secretion of a urine which has a strong violaceous odour. Taken internally for the various diseases, for the treatment of which it has enjoyed some reputation, it occasionally, in small repeated doses, produces inflammatory engorgement of the kidneys, the symptoms of which are as follows:—The patient becomes sensible of a dull, aching pain in the meatus urinarius, sometimes referred to the glans penis, sometimes to the neck of the bladder; in women it is chiefly referred to the pubis: an urgent desire to pass water accompanies this, and the effort is often abortive, the sense of distress increasing with these ineffectual efforts. If any urine passes, the first drops are either pure blood or urine mixed with blood, and a sense of weakness and prostration, even of nausea, indicates the mischief which the turpentine has caused. The blood-tinged urine throws down a coarse, brownish-red sediment, which, examined by the microscope, reveals a multitude of blood-cells and blood-casts.

The immediate discontinuance of the turpentine is followed in a few hours by a mitigation of the symptoms. The urine increases in quantity, becomes less tinged with blood, and in forty-eight or sixty hours has recovered its natural quality. Remedial measures beyond those of rest and drinking freely of diluents, avoiding all stimuli, are rarely required. A warm bath, or a hip-bath at the outset of the symptoms, may expedite relief.

The symptoms excited by the tincture of cantharides taken internally for medicinal purposes in cases where, from peculiar idiosyncrasy, renal disorder occurs, differ but little from those which turpentine excites. Strangury, bloody urine, and intestinal irritation, comprise the most prominent. Intestinal irritation, with discharges of blood by stool, with distressing tenesmus, prove that cantharides operates over a wider sphere than turpentine, involving the intestines as well as the kidneys in disturbance.

In addition to the above it may be mentioned that in those cases which have come under my observation of the effects of cantharides taken medicinally, a violent pain referred to the pubis, and extending in the male through the penis, followed by most painful priapism, described by one patient as similar to the state called chordee during the inflammatory stage of gonorrhœa. In women this symptom manifests itself in the shape of a painful sense of aching along the meatus urinarius, with general sense of irritability in the vulva and clitoris, and an urgent desire to relieve this by pressure, friction, or some mechanical excitation of the genital organs. But this state, both in the male and female, is often accompanied by a burning sensation in the throat, vomiting, sometimes of blood, with symptoms of tenesmus, and bloody discharges from the bowels. It will be thus seen that the action of cantharides is not confined to the renal and genital organs, but extends widely through the gastro-intestinal canal, so that it can never, in any given case, be predicted of this agent that it will exercise a safe amount of stimulus to the organs of generation without, at the same time, implicating the gastro-intestinal tract of mucous membrane in a hazardous, not to say dangerous degree.

Whenever during employment of cantharides as a medicinal agent any, even the slightest indication of its poisonous action becomes manifest, its use must be immediately discontinued. As I am not considering the treatment of poisoning by Spanish fly beyond that required for symptoms arising from its specific effects on the kidneys when medicinally employed, it remains only to observe that the strangury and distress arising from the suppression of urine is best relieved by warm baths, and drinking freely, if the state of the stomach permits, of bland fluids. Cupping across the loins may be necessary. Opiate and oleaginous clysters should never be omitted; indeed, the relief to the tenesmus which opiate enemata produce is quickly followed by an alleviation of the renal symptoms; and the bad consequences of the remedy when medicinally employed rarely go further.

In one case, of a young woman for whom the tincture had been prescribed for some uterine derangement, and who had doubled the dose, or taken the medicine more frequently than had been ordered, in the hope probably of hastening its effects, the strangury continued for three days with considerable irritability of the stomach and much tenesmus. The urinary secretion was fully re-established by the fifth day. No trace of blood-corpuscles could be found by the microscope; and the urine was quite free from albumen. Blood-corpuscles and fibrinous casts of coagulated fibrine were seen so long as blood was present.

A fatal case of poisoning by cantharidis is recorded by Dr. A. Taylor in his work 'On Poisons.' The kidneys were deeply inflamed. It is to be regretted, as Dr. Taylor remarks, that in the few fatal cases of death from cantharidis, that the post-mortem appearances should not have been more accurately and minutely described.

It has been already remarked in the section of causes of nephritis, that the nitrate of potash in concentrated form has been known to produce strangury and nephritis. The cases on record arose from taking the nitrate in mistake for Epsom or Glauber's salts. Such accidents are of rare occurrence now. In the event of such happening, vomiting should be immediately, or as soon as possible, excited. If this be delayed, and the salt be retained, its action is usually shown by griping and purging, to which may be added vomiting. The action of the nitrate is exhibited rather as a gastro-intestinal irritant, than as exciting nephritis. Copious diluting drinks, a warm bath, and subsequently a free action of the bowels, either spontaneous or obtained by oleaginous enemata, will be the most appropriate means of relief. There appears to prevail, even at the present day, a notion that nitre is capable of producing strangury and nephritis. I believe this to be a mistake. Griping, purging, vomiting with great depression, are the chief symptoms mentioned when from one to two ounces have been taken by mistake, but in none are the symptoms of nephritis mentioned. See Christison 'On Poisons,' Dr. A. Taylor, and Dr. Guy, &c.

The oleo-resin copaiba given in doses beyond half a drachm or a drachm, and repeated through the incaution of the patient, may, in addition to great excitement of the genital organs in the shape of painful priapism, induce scanty urine and renal congestion.

In one case in which I was asked whether the blood contained in the urine of a patient who had imprudently taken double doses of the copaiba mixture came from the kidneys or not, I had no hesitation in saying that the blood was urethral and not renal. I have seen no other example of suspected renal disorder from the action of copaiba.

The effect of this oleo-resin on the urine must not be forgotten. Not only is a balsamic odour imparted to it, but a portion of the resinous principle appears in the urine, and may, by the cloud which nitric acid produces, be mistaken for albumen. The cloud does not settle, however, after being set aside. Heat, moreover, redissolves the turbidity.

The action of the salts of lead, as well as that of phosphorus, is slowly and insidiously to affect the nutrition of many textures and organs. The kidneys become implicated in this widespread influence. Their secreting power becomes deranged, the urine is albuminous, and contains evidence of a fatty and amyloid degeneration.

But neither of these poisonous agents seem capable of exciting acute inflammatory action. It is rather to their influence on the organs of nutrition that we trace those widespread ravages of degeneration which are recognised in the bodies of those who, following certain trades, become the victims of several forms of disease, in which the kidneys, with other organs, are implicated.

The urine of every patient suffering from colica pictoneum

should be examined for albumen. In the great majority of cases, particularly in first attacks, no albumen is found. Afterwards as the constitution of the patient begins to succumb to the widely diffused poison, this substance affords a most material fact for our prognosis as well as treatment. In these cases, the casts accompanying the albuminous urine are usually of the hyaline variety, with a few resplendent or highly refractive granules, the nuclei of disintegrated cells. The most unfavorable opinion must be always entertained of the ultimate condition of such patient. A tonic, nutritive plan of treatment with steel in some form, and small doses of the iodide of potassium in some appropriate mode of combination, offer the best hopes for temporary improvement.

Phosphorus exercises no primary or specific action on the kidneys. The kidneys undergo a rapid species of degeneration, subsidiary to the poisonous action of the phosphorus on the periosteal and bony textures of the teeth and jaws. In patients thus affected the urine becomes albuminous, and the sediment will contain similar evidence of renal degeneration from the presence of hyaline casts both of large and small diameter, with nuclei in greater or less abundance dispersed through them—the débris of broken-up cell structure. In this form of disease palliative means can alone be employed.

§ 4. NEPHRITIS *after scarlet fever.*—Synonyms.—*Dropsy after scarlet fever; acute morbus Brightii; acute albuminuria.* —From ten to fourteen days, rarely exceeding twenty-one days after the subsidence of the eruptive stage of scarlet fever, except in special occasions mentioned hereafter, the convalescence of the patient becomes interrupted by a slight feverishness; in some there is a distinct rigour, in others a mere restlessness with inappetency and some thirst, but with all a marked diminution of the quantity of urine passed, and an equally marked alteration in its appearance and quality. In some patients there will be, for some twenty-four hours, almost total suppression of urine; the little which is passed is like pure blood. In others, the quantity of urine, though scanty, is evidently mixed with blood; it is either red as if from simple admixture of blood, or it is brownish red, with a copious brown deposit from the action of the urine on the constituents of the blood passed with it. In other cases the evidence of the presence of blood to the unaided eye is less apparent. The urine not much deficient in quantity, has a dirty smoky appearance, and the sediment is of dark colour, as if a little soot had accidentally got mixed. In another class of cases the urine is clear, but of a peculiar greenish hue, and lastly, the urine may, to the unaided eye, appear natural, but when examined chemically and microscopically, there is abundant evidence of its partaking, though in less degree, of the qualities of all the above-mentioned varieties. They all contain blood in various degrees or proportions, and consequently all are albuminous. The character of the urine, microscopically considered, will be discussed hereafter. In addition to these alterations in the quantity and quality of the urine, a most striking alteration in the aspect of the patient is obvious. There is a remarkable alabaster whiteness of the skin. The features are swollen, the eyelids tumid and infiltrated with fluid; the whole surface of the body is œdematous, and deeply pits on pressure. The trunk, arms, hands, legs, and feet are equally dropsical, the fluid not predominating more in one part than in another. A similar dropsical condition prevails in internal parts. Dyspnœa and wheezing testify to respiratory difficulty from pulmonary œdema. The pulse is quick, exceeding 100; and there is great irritability in the action of the heart; the peritoneal cavity contains fluid. The patient is listless, may be comatose from serous infiltration within the cranium. Such are the more prominent symptoms diagnostic of the renal disturbance which occurs after scarlet fever.

There are other conditions worthy of note not so obvious to the superficial observer. The qualities of the blood are remarkable—deficient in fibrine as well as globules, the serous element largely predominating; there is before us a watery blood, called hydræmic, sufficient to account for the irritable action of the heart, for the obstructive condition of the capillary circulation, and consequently for the diffuse dropsical effusion. Blood deficient in any of its elementary constituents, deficient in globules and surcharged with serum, passes with difficulty through the capillary vessels. The heart labours to overcome the retarded flow, but the stimulus conveyed to it is by an imperfect fluid—its irritability is increased, and that peculiar thrill in its action is felt, which is so easily recognised in this form of renal dropsy, as also in those of a similar type which are unconnected with scarlet fever.

At the outset of this disturbance the urine is scanty, dark coloured; bloody in various degrees; of high specific gravity, from 1024 to 1030, and highly albuminous; the sediment contains many objects which, when examined by the microscope, are signally diagnostic of the character of the renal disorder. Blood-corpuscles in great abundance; coarse fibrinous casts containing blood-corpuscles (blood-casts); and even thus early large coarse granular casts containing a few epithelial cells (granular epithelial casts). These objects, slowly, as the hæmorrhagic stage, or that of stasis, passes away, give place to granular epithelial casts; epithelial cells, in varying forms of development, are seen either free or isolated, or impacted in the larger form of cast. There are multi-nuclear cells (compound granule cells), and associated often with an abundance of free nuclei, these latter appearing as highly refractive granules diffused through the cast.

The casts, as the case proceeds, become less granular, more transparent, with fewer epithelial cells, but with an increase of the abortive cells (granule cells); these are seen as large cells filled with fat granules, highly refractive. The casts often contain these in abundance, together with a great number of highly refractive granules quite free. They result from the disintegration or breaking up of the wall of the large compound granule cells. These objects seen day by day are of most unpromising import. If they decrease in number, if the casts become less granular, more hyaline, with only here and there a nucleus of a broken-up cell, this will correspond with a general improvement of the prospects of the patient; a diminution in the amount of albumen, and a decrease, probably a complete disappearance, of the dropsy. The patient's appetite improves, micturition becomes less frequent, the urine of a healthy quantity, and the sediment only now and then contains a delicate transparent cast, with a few isolated granules. The urine, however, does not quickly lose the albumen; it may continue albuminous long after any objects derived from the kidney can be detected by the microscope. The health and vigour of the patient is slowly regained, and in the course of some three or four months not a trace of albumen can be detected in the urine. While in others, notwithstanding the return of good health, albumen continues present for two or three years after the date of the attack of scarlet fever.

The presence of albumen in the urine is therefore not necessarily a dangerous symptom, though of serious significance. It must be taken in connection with the antecedents of the patient as well as with the objects which may be detected in the sediment by the microscope. I know personally several men in the prime of life, healthy and vigorous, and fathers of families, who, when children, had scarlatinal dropsy, with albuminous urine continuing, in one case, for eighteen months, in a second, beyond two years, in the third it was not till the full period of puberty had passed, and early manhood was developed, that the albumen, present for five years, at length totally disappeared. Diphtheria will leave behind a disposition most inveterate to the continuance of albumen in the urine; the health remaining unimpaired notwithstanding the albuminous state of the urine. It may be stated that in these cases of persistent albuminous urine after scarlet fever and diphtheria, the albumen undergoes some modification in its ordinary character, and may be overlooked by those not conversant with the fact, that in a certain stage of these cases, the albumen becomes modified, and acquires the character of what the chemist recognises as the deutoxide of albumen.*

* See Part III, Albuminous Urine.

When the progress of a case of scarlatinal dropsy is of an unfavorable character death may take place within the first week; or the fatal result may be prolonged for weeks, or even months.

In those cases where a fatal issue rapidly follows the development of the renal symptoms, the immediate cause is either some affection of the nervous system arising from uræmia, or the rapid effusion of a bronchial fluid into the air-passages, and death from apnœa.

In the first class of cases the symptoms referable to the nerve-centres may either be a profound coma or a series of convulsive attacks terminating in coma.

In the more protracted cases it is not possible to predict of any given case that it will not be carried off by symptoms of uræmia. It is well to prepare those interested in the patient by the possible development of such symptoms; though most of these cases live through many months without any indication of such a termination.

In these protracted cases ultimately proving fatal, it has been already remarked that the microscopic objects continue to exhibit a persistent disorganization of the cell-structure of the kidney. The urine, however, increases much in quantity, becomes pale, of a much lower specific character, may be as low as 1006-1008, without any diminution in the quantity or quality of the albumen. The termination of these protracted cases may be either by some affection of the nervous system from uræmia, from effusion into the bronchial tube, or serous exudation into the pericardium; various complications may, however, arise in these cases beyond those just mentioned; but the above are by far the most frequent cause of the fatal result.

§ 5. SCARLATINAL DROPSY.—*Treatment.*—The principles on which the treatment of these forms of renal disturbance is based are these—

I. To restore, if possible, the equilibrium of the circulation through the kidneys.

II. To lessen the amount of the dropsical effusion, and thus relieve the lungs and other organs.

III. To aid the nutritive functions, and thus promote the restoration to the blood of those qualities on which healthy cell-development must depend.

The febrile excitement, usually not very great, which marks the commencement of this secondary disturbance is best relieved by some ordinary diaphoretic saline, cooling drinks, and brisk purgatives. The citrate or acetate of ammonia, with some other neutral saline, or the carbonate of ammonia in a state of effervescence, with prepared lime or lemon juice, or the juice of the fresh fruit combined with nitric ether with spirits of chloroform, are appropriate means of obtaining this end.

By the action of a brisk purgative the congested kidneys are also relieved. The diet should be nutritious, but not stimulating. Beef-tea, veal broth, chicken broth, or in very young children, a purely milk diet will be suitable.

The gorged kidneys may be further relieved by hot-air baths—the warm bath is not so efficacious, but it answers sufficiently well in children better than in adults suffering from acute dropsy.

When the urine is highly loaded with blood and very scanty, dry cupping across the loins, to the extent of two or four cups, will help to restore the flow of urine, which returns as the engorged kidneys are relieved. The presence of blood is not the indication for dry cupping—it is the scanty amount, the all but suppression of the urine, which is the significant symptom.

In children, as well as in adults, the clothing even in bed should be flannel during this period of the disorder. The subsidence of the feverishness which rarely lasts beyond forty-eight or sixty hours, is marked also, in the great majority of cases, by a palpable diminution of blood in the urine; and a marked increase in quantity. If the bowels have been acting well, there is also a decrease in the dropsy not only of the surface but of the internal organs, the breathing is relieved, the pulse less frequent, and the action of the heart less irritable. Further, to lessen the dropsical effusion, and to give relief to the kidneys, brisk and active purging for a few days should be kept up, and the best agent, both in children and adults, is a combination of jalap, cream of tartar, and ginger.

The Pulvis jalapæ compositus of the Pharmacopœia, according to age, ten, twenty, thirty, or even sixty grains, may be given daily or on alternate days, simply mixed with water. In the morning, fasting, is the best time for administration. Children rarely require stronger purgatives. Elaterium, podophyllin, &c., may be given to adults.

The dropsical effusion slowly declines, the quality of the urine improves, and usually within the first week, in some cases even earlier, the third object of treatment may be commenced. A simple, but well-regulated diet should be enforced; all pamperings of a capricious appetite strictly forbidden; a definite number of meals at stated fixed hours should be prescribed, with a caution that no *sweeties* in the intervals should be given the child. Everything should be done to promote healthy and vigorous digestion, and every thing avoided calculated to lessen or impair it.

It has been already shown in the pathology of this form of nephritis what destruction of blood-corpuscles has taken place, how deficient in these the blood has become, and how this poverty of corpuscles may be considered as a guide for our future treatment.

To restore this deficiency is embraced in the third principle of treatment.

Preparations of iron are the best aid to the blood-forming function. But iron in any of its preparations cannot alone generate blood-corpuscles, they can only be formed out of the elements of nutrition found in food; but it is remarkable how much more speedily convalescence proceeds, how quickly the pallid wax-like aspect is exchanged for a more healthy hue, when iron in some appropriate form is given in conjunction with food. Replacement of those cells in the tubular structure of the kidney, which have been disintegrated and removed, as the microscope has shown in the acme of the disorder, is the final end in view. And as this renewal of secreting cell-structure proceeds the albumen decreases and ultimately altogether disappears from the urine.

Of the many preparations of iron, the perchloride is the most effective in this disorder. It may be given in a few drop doses, with simple syrup. It may be given in similar small doses, five to ten drops, in seltzer or soda-water, and thus a factitious chalybeate be produced. A long experience of these and other forms of renal disease where the object of treatment is similar has, however, convinced me that a soluble ammonio-chloride, obtained by acidulating the Liquor ammoniæ acetatis with dilute acetic acid, and then adding the tincture of the perchloride, is the most efficacious of all the so-called preparations of steel. This form may be prescribed with whatever dose of the perchloride it may be thought desirable to administer; but the preparation must be made up in the manner described, or an insoluble sediment forms.

In very young children, or in patients up to eleven or twelve years of age, wine is not needed, provided food is taken without reluctance. In those in whom the appetite is fastidious, a very small quantity of port wine taken with the food of the chief meal may be found beneficial.

Among the other preparations of iron in the Pharmacopœia, adapted for use with children, is the syrup of the phosphate of iron, the Vinum ferri citratis, and the ammonio-citrate.

Little more can be said upon the treatment of those cases, in which, even from the first, the eye of experience sees the prospect of an unfavorable termination. The same principle must govern both types of cases. Symptoms of uræmic poisoning are oftentimes so quickly followed by fatal termination, that little can be done with any prospect of successful relief. Active hydragogue purgatives are alone to be relied on.

Chlorine given off by the addition of one drachm of strong hydrochloric acid to ten grains of the chlorate of potass in a ten-ounce stoppered bottle, and the bottle filled ounce by ounce, shaking on each addition, may be given, but it is less effective than in the primary stage of scarlet fever, particularly in that form known as scarlatina anginosa, when the antiseptic properties of chlorine exercise a very beneficial effect on the foul exudations of the mouth and throat.

In those cases in which an albuminous urine remains after the renal dropsy and other symptoms of kidney disturbance have disappeared, it must be stated that this condition of the urine must be prudently considered as a matter of grave importance. If the general health be maintained, and there be this morbid symptom alone left, the patient should be placed on his guard, warned as to the urgent need of avoiding all circumstances calculated to interfere with the health, for while that state of urine prevails, although it may eventually disappear, there can be no certainty offered that the individual is safe from the ulterior consequences of chronic disease of the kidney. It is true we meet with many cases in which a protracted period of albuminous urine is not followed by permanent disease of these organs. But it can never be predicted of any individual case that it will with certainty be included in that category.

§ 6. NEPHRITIS *from cold and wet; probable predisposing causes,—the strumous or scrofulous constitution; habits of intemperance; syphilis.*—Synonyms.—*Acute morbus Brightii; acute renal dropsy; albuminous nephritis; inflammatory albuminuria; acute desquamative nephritis.*

Symptoms.—This form of renal disease may occur at any period of life, but it is more prone to be developed in the early period in those of strumous, rather than in more energetic constitutions; but in the middle period a marked predisposing cause appears to be excess in, or addiction to, alcoholic stimuli, but there the form of disease is usually chronic rather than acute. The early stage is strongly marked by evidence of inflammatory engorgement of the kidneys, primarily, and of the lungs, pericardium, peritoneum, secondarily, or in the course of the disorder.

A state of feverish excitement, hot skin, rapid pulse, and

dyspnœa, quickly followed by a sense of exhaustion and bodily prostration mark the general disturbance, while nausea and retching, with scanty and all but suppressed urine, which is dark coloured and bloody, and of high specific gravity, indicate distinctly enough the share the kidneys have in the disorder. A puffy, œdematous state of the eyelids and cheeks will often appear during the first twelve hours—usually in the morning of the first night. The pulse is quick, sharp, and hard, heart's action irritable, just as observed in scarlatinal dropsy. There is thirst, inappetency, more or less distress in the breathing, not dependent on bronchitis, but on a dropsical state of the pulmonary parenchyma. Within twenty-four hours the whole surface of the body will become anasarcous; trunk, arms, legs, and feet, all pit deeply on pressure; the cuticle acquires, even in adults, the same remarkable alabaster paleness that was noticed in scarlatinal dropsy. The state of the blood is the same—watery, the serum of low specific gravity, the blood-corpuscles deficient; and the sum of the symptoms similar in every respect to what has been described in the renal disturbance succeeding to scarlet fever. Serous effusions may take place into the pulmonary parenchyma, into the pleural sac, into the pericardium, into the peritoneal cavity, or poured out from the arachnoid, and symptoms of coma may come on from conditions other than uræmic, from pressure of the effused fluid on the cerebral hemispheres; yet these conditions are more frequently observed in the chronic form of the disease.

The sum of the symptoms being the same, the pathological conditions identical, it follows that the treatment of this form of renal inflammation will differ in nothing from the former. Hot-air baths; dry cupping from the loins; diaphoretic salines; hydragogue purgatives, and subsequently ferruginous tonics, of which the soluble ammonio-chloride is the best. The diet should be nutritious, light, care being taken that the digestive function is not oppressed by too bulky a meal. The need for stimulants is greater, however, than in the form described as following scarlet fever. Not from any distinction in the character of the disease, but that adults require stimuli, as a rule, more frequently than patients during childhood or before puberty.

The probability of permanent recovery from renal dropsy arising from causes other than scarlet fever, speaking with the light of experience, is more doubtful. Much depends on the age, the antecedent health, the family history, the habits of life of the individual.

Each case, in respect of prognosis, must stand by itself, for certainly many cases where the disorder has come under treatment early have permanently recovered, yet under any aspect the malady is not lightly to be dealt with, nor viewed otherwise than seriously damaging the expectation of life. It is worthy of remark, however, that as a rule, acute cases are far more amenable to treatment than chronic ones, and therefore the prognosis of any given acute case will depend very materially on the period of the disorder at which the treatment commenced. If within the first few days, a pretty confident expectation of recovery may be entertained, and this is the reason why scarlatinal nephritis is so successfully treated.

The following case may be quoted, it having happened recently, of the favorable termination of a case of acute morbus Brightii, notwithstanding the existence of a strongly pronounced strumous habit of body; the patient coming under treatment three days after the first indication of the disorder:

J. C—, a lad of seventeen, a printer's boy, of a strumous diathesis, having enlarged submaxillary glands with scars on both sides of the neck, noticed that a day or two before admission his face became swollen, the day previous he had passed but little water, and very frequently, and what passed was like blood. In the following afternoon his legs swelled; the abdomen and scrotum also became distended. He was admitted the day after to the Westminster Hospital. General anasarca pervaded the whole body. The face and eyelids were swollen; there was a marked waxy appearance of the skin. The arms, hands, thighs, legs, and ankles pitted deeply on pressure. The scrotum was infiltrated with fluid, and the abdomen was distended, but no distinct fluctuation was detected. There was a good deal of dyspnœa, coarse wheezing murmurs were heard in the chest. There was no cough; the pulse was 90, and no elevation of temperature. The urine was clear, with a distinct sediment, in which isolated blood-corpuscles and some fibrinous casts were seen; it contained a large quantity of coarsely coagulating albumen. He was ordered the compound jalap purgative, warm bath, and subsequently the usual ferruginous tonic. The urine daily increased in quantity, and the anasarca of the body as proportionately decreased. Before the tenth day all dropsical conditions had disappeared. The breathing was quite free. He required the purgative on alternate days. On the third week the urine contained but a trace of albumen, and a few hyaline casts were alone visible to the microscope; at the end of the fourth week he was discharged convalescent. No cause was assigned for the commencement of this attack; he had not been exposed to cold or wet; no scarlet fever existed in the house or place he frequented. But he was unequivocally of a well-marked strumous habit.

Though the exciting cause cannot be traced, the predisposing cause was strongly apparent. But strongly marked as it was, and unpromising as this habit of body is when assailed by renal disorder, yet fortunately commenced in the earliest period of its manifestation, four weeks were sufficient to establish a convalescence. To keep this lad under inspection for a few years, would add much to the interest and value of his case.

Acute morbus Brightii.—Treatment.—The general principles already stated upon which the renal disease, and dropsy after scarlet fever should be treated, are equally applicable to cases of acute albuminuria occurring from other causes.

If the temperature of the body is above the natural standard, simple diaphoretic medicines, such as the ammonia salts in a state of effervescence, with brisk purging; the warm or hot-air bath, and, in some cases, dry cupping across the loins, will be the most suitable treatment. So soon as the feverish heat abates, while the purgative effects of such combinations as the compound jalap powder of the Pharmacopœia may from time to time be needed, the patient should be placed on some ferruginous tonic, with such an improved scale of diet as the digestive organs can bear. Stimulants in the form of wine or beer may be given, indeed, in some cases, from previous habit of the individual, are absolutely needed. I will repeat here a caution in the use of stimulants in these cases which I have made elsewhere, and which an extended experience amply confirms, that they should never be taken without food. With food, or immediately after, they promote digestion, and invigorate, taken on an empty organ they hurriedly and immediately pass into the circulation, excite the heart, temporarily elevate the temperature of the body, and convey a very hurtful stimulus to the kidneys. In no disease is this warning more necessary than in acute albuminuria. See also under the section Chronic Albuminuria.

§ 7. RHEUMATIC NEPHRITIS.—*Symptoms.*—The symptoms of rheumatic nephritis will require but a very few words. It has been already remarked under the head of the causes and pathology of rheumatic nephritis, that the occurrence of this complication in acute rheumatic fever is of rare occurrence. Its very existence during life may not be suspected, and that for the most part it is developed during the last few days of life, and consists either in embolism of the renal vessels, or a spontaneous exudation of fibrine within the wall of the vessels, and a consequent plugging up of the channel of the renal circulation.* Life is rarely prolonged sufficiently for any symptoms of a prominent character referrible to the kidneys to be developed.

§ 8. GOUTY NEPHRITIS.—*Nephralgia; lithiasis or gravel; stone in the kidney.*—In treating of the causes of gouty nephritis, its connection with, even its dependence on, the uric acid accumulation and deposit in the kidney was distinctly stated. The connection, moreover, of gout with renal calculus or stone in the kidney, has been the subject of observation from the earliest period of medicine. The symptoms of nephralgia, lithiasis or gravel, and stone in the kidney, are so intimately associated one with the other, and as they may be considered as diagnostic of conditions all running towards the same point, practically, it may be of advantage to take them together.

Nephralgia; gravel; lithiasis.—Nephralgia simply implies a pain or series of painful sensations accompanying gravel. Strictly speaking, these morbid conditions are not dependent on inflammatory action, and they appear misplaced in the group; but hæmaturia is a frequent accompanying symptom. Any kind of pain across the lumbar region may at random be called nephralgia, under the notion that the origin of it is in the kidneys, for the term implies pain in the kidneys. But unless the pain or uneasiness be accompanied by some special circumstances, the term is misapplied.

The character of the pain in nephralgia is almost as various as the patients who suffer. Sometimes it is pain like that in lumbago—only felt on movement, but it is attended by a peculiar aching referred to the crest of the ilium. The pain may extend from the lumbar space on one side or the other down the side of the abdomen, following apparently the course of the ureter. There may be curious sensations, difficult to describe, located in the large cushion of the glutei muscles of the buttocks; one patient spoke of it as if the muscular bundles were strained and tied up. Uneasy sensations, spoken of as rheumatic, were felt in the large muscles of the thigh; there may be some defective sensation, a mere numbness of the external part of the thigh. These symptoms are invariably accompanied by increased frequency of micturition, particularly at night. The urine has no morbid character visible to ordinary observation, but when chemically estimated, uric acid is found in it largely in excess, and frequently forms as red or orange-red crystalline grains on the urine cooling, sparsely on the inside of the chamber vessel. If such urine be allowed to cool in a glass vessel, these red grains are easily distinguished on its sides. The urine may form in still greater excess, and a distinct sediment of red sand, in considerable abundance, may be found on the urine cooling, or a distinct crystalline deposit. This nephralgia, with its accompaniment of uric acid red sand, is rarely associated with any marked deviation of the general health, it may disappear spontaneously, and it may return on any dietetic imprudence. It may thus return at intervals for years, and it may eventually disappear altogether, but in the great majority of cases, these indications of the gouty diathesis or tendency, are but the precursors of a genuine attack of gravel.

GRAVEL.—In this, all the sympathetic symptoms just mentioned may be present, but not invariably. The most uniform diagnostic group of symptoms consists in a peculiar aching pain, sometimes described as darting and pungent, referred to the neck of the bladder and extremity of the meatus urinarius; with this is associated a constant irrepressible desire to pass water, with only a temporary relief to the pain at the extremity of the urinary passage; very little urine passes each time, and its course along the urethra is marked by a burning cutting pain, as if the passage was denuded of its protective epithelium. In mild cases the urine is not turbid, although, on settling, a coarse sediment falls of a fawn colour, and composed of particles of greater or less size, like coarse sand, or even gravel.

In other cases, after a period of longer or shorter duration, in which the frequency of micturition is the most prominent symptom, the patient, if it be the first attack, is alarmed at the appearance of blood in his urine, and when the urine has settled, the sediment is found to consist of a layer of blood globules, among which the same kind of sandy or gravelly particles just described may be found. It is worthy of recollection that in these cases of hæmaturia accompanying gravel, the blood is never passed in coagula. The gritty sediment when examined is composed of almost pure uric acid. The duration of these attacks of gravel, as well as that of the hæmaturia accompanying them, is usually very short, lasting rarely more than forty-eight or sixty hours. The urine becomes clear and natural looking, and contains for a few days some scattered blood discs, crystals of uric acid, and, perhaps, a few exudation or mucous corpuscles. Now, all these symptoms may occur to a patient who never had gout. They, however, strongly indicate the gouty constitution, and imply either that some day or other the individual may have gout, or that the elements of renal calculus exist, and the probability that the latter disorder will eventually declare itself.

It has been remarked that these symptoms of gravel, with or without hæmaturia, may occur to a person who never had gout. They, however, as often occur to the sufferer from that disease, materially aggravate his sufferings, and offer a far more unfavorable prognosis than when the individual is free from those renal complications.

§ 9. GOUTY NEPHRITIS.*—It is very doubtful whether the kidneys are subject in the gouty constitution to inflammatory action, apart from what may be excited by gravel or the lodgment of a calculus within them. The stomach, the intestines, the lungs, the heart, the bladder, even the brain, may, in the retrocedent form of gout, be seriously implicated; symptoms of nephritis, if hæmaturia be taken only as the index symptom, are also of frequent occurrence; but then these renal symptoms are, so far as my observation enables me to judge, invariably excited either by gravel or by a calculus in the kidney, so that it is not only difficult, but impossible, to distinguish between gouty nephritis and calculous nephritis occurring in a gouty patient. Indeed, they may be taken as convertible terms. Gouty nephritis is caused by the escape and irritation of calculus and gravelly matter in a gouty patient,

---

* The reader is cautioned against confounding this form of renal disease in gout with what Dr. Todd called atrophic or gouty kidney, and which will be treated of under that head.

while calculous nephritis may occur to one who never had gout, the symptoms of which are obviously similar and arise from the dislodgment of a minute solid concretion, which, up to the time of its displacement, had remained inert and unknown.

The intimate connection between gout and renal calculus has been already noted. Gouty nephritis may be recognised by the following symptoms:—In the interval of the gouty attacks, and after some days of considerable irritation about the urinary passages, shown by frequency of micturition and aching in the meatus and uneasy pains across the loins, the urine being for the most part pale and clear, urgent and painful desire to pass water comes on every ten minutes or quarter of an hour, with straining and ineffectual efforts to empty the bladder, the urine passes only drop by drop, and presently blood is voided. The stomach becomes irritable, nausea and a disinclination for food has probably preceded these symptoms; these increase with the renal distress; urgent pain occurs in one side or emanates from one kidney; a pungent, colic-like pain darts from this side down the direction of the ureter, and the straining and ineffectual efforts to empty the bladder or to relieve the sensation of its fulness is followed by retching and vomiting; cold sweats break out from the head and forehead, which soon pervade the whole surface; the pulse becomes small and weak; a general sense of prostration and exhaustion is expressed in the features of the patient, who continues probably an hour or two in this state of suffering; more blood passes, and probably with it some grit or gravel. The painful desire to pass water decreases, more urine flows, the stomach becomes quiet, retains what is swallowed, and a general sense of relief succeeds. For some days, however, the patient requires quiet and bodily rest. The urine, for many days after, contains uric acid grit with isolated blood discs and an abundant sediment of crystals of uric acid. These symptoms differ but little from an attack of gravel; indeed, it is an attack of gravel in a gouty patient, but the symptoms are more severe, the prostration greater, the depression of the heart's action more obvious, and the succeeding febrile reaction more marked. Again, the symptoms are, in many respects, similar to what is called renal colic occurring in calculous nephritis from the escape of a stone and its passage down the ureter. The griping colic-like pain in this so-called gouty nephritis arises from the same cause—the passage of gravel from the kidney to the bladder—but it must be noted that this colic-like pain is not present in every case; it is not, therefore an essential or typical symptom. The half-suppressed scanty urine, the painful frequency of micturition, the hæmaturia, the exhaustion, the retching, possibly vomiting, are all typical and diagnostic of this form of nephritis.

The treatment of these cases will, to a certain extent, vary according to the prominence or absence of particular symptoms, or idiosyncracy of the individual. A warm bath, a brisk purge if the bowels need, cupping across the loins, the moderate employment of narcotics, and perfect rest in the horizontal position will be required in every case. As soon as the urine exhibits less blood, the patient should drink plentifully of diluents; stimulants should be sparingly given, and chiefly to those whose previous habits render them necessary.

The sequel and termination of these cases is usually in either calculus or stone in the kidney, or more frequently in calculous pyelitis, the symptoms of which will be presently detailed.

Nevertheless cases of gravel and hæmaturia do occur which are deficient in the diagnostic symptoms above mentioned. I was requested to see a gentleman in his eighty-eighth year, who suffered much from urgent frequency of micturition during the day. At night he was not disturbed. The urine was examined chemically, and the sediment by the microscope. A trifling increase in the amount of uric acid, and the presence of numerous crystals of oxalate of lime on the urine cooling was evidence of an excess of that kind in the urine. These crystals were seen in unusual abundance in the sediment, and with them many epithelial cells from the pelvis of the kidney and ureters. The carbonates of potash and lithia in a state of effervescence with lemon juice were prescribed with great advantage to the only symptom that distressed the patient—the frequency of micturition. The disorder was recognised as probably due either to a threatened attack of gravel or the earlier symptoms of renal calculus. This gentleman was of hale and hearty constitution, had never had gout nor any other illness. The only disorder which had troubled him was an affection of the skin, which I noticed had been a form of eczema. This cutaneous disorder, as well as the scaly varieties of lepra or psoriasis are, in all constitutions, but more particularly in the aged, associated with an excess of uric acid, or, in other words, with the gouty element.

A year passed, and in the spring the annoyance of a frequent need to pass water returned. There was no urgency during the night, the power of retention in the bladder during the hours of rest was apparently natural. There was not sufficient prostatic enlargement to account for the symptoms during the day, and any disease there would have manifested itself during night as well as day. In the beginning of summer himself, as well as his medical attendant, were alarmed at a pretty copious hæmorrhage occurring each time he passed water. There were no clots; it was liquid blood mixed with urine. There were no subjective symptoms of a typical character, nothing but the hæmaturia and the frequency of passing urine. On the second day of the hæmaturia I saw him; there had been, in the morning, a difficulty in passing his water, which was momentary, and he was sensible of something solid having passed; twice in the course of two hours this occurred. When I examined the vessel in which the bloody urine had been passed, there was a large deposit of a coarse gravel, fawn-coloured when washed, and there were several large aggregations of the same material, so large as to be an object of wonder how they could have passed unaided through the urethra. The quantity collected and carried away for examination exceeded thirty grains, and proved, when examined by the ordinary methods, to be perfectly pure uric acid, with a minute trace of oxalate of lime, as shown by the ash left before the blowpipe. The potash salts were again given, complete rest enjoined, and the drinking freely of any liquid capable of promoting diuresis—hock or claret with Seltzer water. The cream of tartar drink, or cold spring water in tumblers two or three times a day, were suggested to be taken according to the taste of the patient.

In three days' time the urine acquired a natural appearance. Here and there a stray blood-corpuscle was visible by the microscope; some oxalate of lime crystals, and a few epithelial cells from the renal cavities were also seen. The general health, appetite, and sleep were unaffected throughout.

This case then illustrates the remark just offered, that many cases, the nature of which the sequel unequivocally proves, occur without the strongly marked diagnostic symptoms so prominently present in others.

In cases where the pain continues unrelieved, anodynes are useful. The compound tincture of camphor is the best calculated to afford relief, as from the benzoic acid which it contains it is calculated to aid in bringing the uric acid into a state more easily adapted for excretion. Hip-baths, or a warm bath, when it can be had, are useful adjuncts to treatment.

§ 12. CALCULOUS NEPHRITIS—*stone in the kidney*. CALCULOUS PYELITIS.—These two forms of renal disorder run so insensibly into one another that it would be an unnecessary piece of refinement to separate any description of their symptoms.

Stone in the kidney is less fatal, less painful, but of more frequent occurrence than stone in the bladder. It distresses the patient less, that is, in the majority of cases, but it is less within the reach of relief. While it remains impacted in the structure of the kidney, or detained in the calyces or pelvis of the kidney, it is clearly beyond the reach of medicinal agents, either for its removal or displacement.

Its further growth or enlargement may be checked, and this, while it remains in the kidney, is all that treatment can hope to accomplish.

Experience justifies the announcement that this result is, in a majority of cases, certainly attained. Individuals having stone in the kidney may live for years, and often attain a fair average age, if their diet and habits of life be properly regulated, and a plan of treatment to be hereafter described steadily persevered in.

The symptoms of stone in the kidney are oftentimes spread over a long interval of time, that is, the disease is of very slow development, and may exist for years without giving any sign, or but very doubtful ones, of its origin. The term nephritis is here certainly misapplied; it may be doubted whether the presence of a calculus ever does more than, when disturbed, excite some irritation, with an escape of blood. The whole organ is certainly, from this cause, never the seat of diffuse inflammation. Hæmaturia occurs with more or less local pain, but never accompanied by any such constitutional disturbance as would be surely manifested if so important an organ were the seat of inflammation, and which we witness in cases of nephritis from other causes.

In many patients there is not even any marked local uneasiness in the loins, and the only symptom the patient can give is, that he passes blood every now and then, once a fortnight, or not so frequent, perhaps. The general health is unimpaired, the urine in the intervals perfectly clear, and it is only by a microscopic examination of the sediment that the nature of the case, is made out. Doubtless, cases with such negative symptoms, are not frequent, but they do occur, and to remove the difficulty of the diagnosis without the ordinary typical symptoms, is the purpose I have in noticing them. In the case mentioned under the head of nephritis from shock or wrench, p. 59, inquiry into the circumstances of the first attack of hæmaturia revealed the fact that the patient attributed the appearance of blood at the commencement to some wrench of the back, or jar, which gave his kidneys a twist, as he thought.

The occurrence of the hæmaturia is thus often found to follow some physical effort or bodily shake, riding on horseback or jumping, or a jolting in a carriage over rough ground. Lastly, the urine, either at the time of the escape of blood, or a day or so after, will present numerous little flocculent shreds, looking, to the unaided eye, like membranous films, but which the microscope shows are collections of exudation corpuscles, entangled within which are seen either amorphous or crystalline grains of uric acid. Here and there a blood disc is visible. The urine at the time may be now and then faintly albuminous, dependent on the presence of the serous element of the blood in small quantity.

The symptoms of the early stage of renal calculus are not always so obscure. Numerous sympathetic sensations arising from the intricate connection of the renal plexus with the numerous ganglia that lie surrounding the kidneys; the abundance of these ganglia on and about the ureters, and the ramification of their branches with those of ordinary cutaneous sensation on the hips and thighs, render intelligible the cause of the pain or uneasiness referred to the anterior or posterior crest of the ilium; the numbness affecting the external crural cutaneous nerve, the drawing up of the testicle by the contraction of the cremaster, the aching at the extremity of the urethra, and other disordered sensations, indicate renal irritation from the presence of a calculus.

The calculus may remain in its seat, giving rise to these occasional attacks of hæmaturia, when, by some accident, it gets at once dislodged, and passes into the infundibulum, and thence down the ureter into the bladder, and eventually, when small, may pass through the urethra, and thus finally escape, freeing the patient from all further suffering or uneasiness. But this event may be marked by typical symptoms of great force, or it may occur without any, but the slightest evidence that such has happened. The dislodgment of the calculus sometimes is marked by the following symptom:—A sense of nausea, followed by vomiting, co-existent with a darting pungent pain from the region of one kidney, and quickly followed by intense abdominal agony, like colic. The vomiting and retching alternate with these agonising gripes, cold sweats break out from the forehead, and, for a while, the colic continues. The duration of these symptoms lasts while the calculus passes down the ureter. They suddenly cease with a sense of immediate and complete relief. The patient, if not cautioned before, should now be told to pass his water always into a chamber-pot, if possible, and to be on the constant watch for any substance passing along the urethra.

A patient may be passing water with his ordinary care when the stream suddenly stops. He strains, and the effort to continue the stream is abortive; the forcing, straining effort continues, and suddenly, with force, out shoots the obstacle to the emptying the bladder, and the distress of the patient is at an end. Instances occur, however, of a renal calculus being thus passed, the pre-existence of which was not even suspected. No renal symptoms of hæmaturia, or of the passage of a calculus down the ureter, having ever been noticed.

The duration of the period of hæmaturia, with occasional local symptoms of pain or uneasiness, is very various in different individuals, and I believe greatly depends on the care the patient takes, and the treatment adopted. In fact, this period depends on the growth or otherwise of the calculus. If that be brought to a stand, a month, even a year or more, may pass without any other symptom than occasional recurrence of hæmaturia, with the little membranous films already described. In very rare cases the disease may stop here. The calculus gets enclosed or shut up, a capsule forms round it, and there it may remain inert for the remainder of life. But the progress of the disease is usually very different. The appearance of the urine slowly undergoes a change. In the intervals of the hæmaturia the sediment, after the urine has been set at rest, will not exceed that of healthy urine, but soon this sediment palpably increases in quantity; and with it the frequency of micturition; with an aching at the extremity of the meatus, relieved by passing even a small quantity of urine.

The sediment when examined with the microscope will show abundance of so-called mucous cells, spherical corpuscles with reniform nuclei, a few multi-nuclear, with here and there an isolated blood disc.

The proportion of sediment is up to this time not so great as to alter the appearance of the urine till it has settled; but soon, the urine, even when passed, appears milky; and when set at rest separates into two distinct portions. The lower stratum has a less flocculent character than the mucous cloud first noticed, it separates distinctly like a precipitate. The upper stratum is clear, but slightly albuminous; the lower is distinctly marked by a complete separation of its particles from the fluid above. Sometimes a faint pinkish line may mark the surface of the sediment. This indicates that blood in small quantity is also present. The larger and heavier pus-cells fall first. The lighter blood-corpuscles settle on their upper layer. At this period of the renal disorder a series of very characteristic sympathetic sensations are usually present. Aching pain in the region of one kidney, extending downwards, and easily traced to the course of one of the ureters; a gnawing pain at the crest of the ilium; defective cutaneous sensations on the outside of the thigh; a constant ache or uneasiness referred to the extremity of the meatus urinarius. In men the drawing up of one testicle, and with these there is a distressing frequency of micturition.

Symptoms still more diagnostic of calculous pyelitis may be present. The urine may vary as to the quantity of pus present: one day a great deal, the next a palpable diminution. About this period of the disorder, a distinct sense of weight and fulness is felt by the patient in one kidney. The stomach, perhaps, becomes irritable at the time when the pus in the urine is least; and a careful examination of the region of the kidneys may detect a distinct swelling or enlargement of the affected kidney. There may be days together when the pus is altogether absent from the urine. The suffering of the patient at these times is sensibly increased.

The renal tumour is also more perceptible. Nausea, possibly vomiting, may occur. The pus appears again in the urine, the stomach gets quiet, the renal tumour is less evident, and with the return of the pus the patient experiences great relief. The explanation of these symptoms is easily understood, if the student will examine a preparation of a kidney with a calculus embedded in the pelvis or infundibulum. These calculi are branched, and by the apposition of urates and mixed phosphates mould themselves, as it were, in this cavity. At first, the flow of pus from the calyces is but slightly obstructed, the obstruction however increases by the growth of the calculus. The purulent urine collects behind it. It dilates the cavity of the pelvis; a renal tumour is formed, and the purulent urine is retained until the pressure reaches a point, which temporarily dilates the head of the ureter, and the purulent urine flows again into the bladder. This happens again and again; the kidney in which this process is established eventually becoming converted into a hollow sac, with complete obliteration of all traces of renal structure, the capsule of the kidney, and the cortical substance condensed into a fibrous-like structure forming the walls of the sac. If the opposite kidney is free from calculous disease, and the disorganization confined to one kidney, the patient may live for years and reach an average age.

It remains now to speak of the treatment of these varieties.

CALCULOUS NEPHRITIS.—*Treatment.*—We have regarded excess of uric acid in the urine, red sandy gravel, and stone in the kidney, as arising from similar causes—it may be well, therefore, to consider the treatment of these several disorders collectively—not separately. The essence of the treatment should consist in the endeavour to limit the increase, and to facilitate the elimination of uric acid from the system. These objects and purposes can only be accomplished through the agency of diet, exercise (regimen), and physic. The word physic, perhaps, conveys an erroneous idea that there are certain remedies possessing specific effects or actions and acting as antidotes to these disorders of nutrition; there are none such. Nevertheless there are medicinal agents which certainly indirectly minister to the great object aimed at—the elimination of the uric acid. But to understand their action, the qualities and chemical properties of uric acid must be first studied.

Chemically, we know that out of the body uric acid may be converted into urea and oxalic acid through the agency of an oxidation established by a fermenting material, a high temperature, and an alkali. The direct oxidation of uric acid by peroxide of lead brings about the same results. Wöhler and Frerichs, moreover, found that uric acid injected into the veins or stomachs of animals was oxidized precisely in the same way as in the presence of the peroxide of lead, and converted into urea and oxalic acid.*

Clinical observation abundantly proves that an excess of uric acid in the urine in the shape of red sand may be quickly reduced by the potash salts, the citrate, acetate, tartrate, not by the combination of the uric acid with the potash, as was once supposed, but by the oxidation of the uric acid in the presence of an alkaline salt and its conversion into urea and oxalic acid.

This interpretation is based on the following facts. The urine of a patient was selected from which a very abundant deposit of red sand (crystalline grains of uric acid) fell down as a copious sediment as the urine passed direct from the bladder. The proportion of uric acid obtained by precipitation from the urine when cold exceeded by more than double the average proportions of uric acid in 1000 parts. The urine had a specific gravity of 1022, and 1000 grs. of weighed urine gave nearly two grains of uric acid obtained by precipitation of hydrochloric acid. The amount of urea in a like quantity of weighed urine was estimated as 36 grs. It must be remembered that the urine had already parted with a large proportion of uric acid in the form of crystalline sediment, so that the proportion of uric acid secreted by the kidneys must have been largely in excess of two parts in a thousand. The

* Lehman, 'Phys. Chemistry,' vol. ii, 453; vol. iii, 417.

patient was placed on half-drachm doses of the carbonate of potash in effervesence with lemon juice every four hours. Three days afterwards, the patient having taken four doses daily, the urine was again examined; no red sand was present, nor was there any precipitation of uric acid spontaneously as the urine cooled. The specific gravity of the urine was 1024. The amount of uric acid in 1000 grs. was estimated at a fraction above a grain. The urea of 1000 grs. was calculated at 48 grs. And there was sufficient present to obtain, in a shallow porcelain vessel, an abundant crop of crystals of nitrate of urea, by adding strong nitric acid to a layer of the urine. A fine flocculent sediment, formed on the urine on cooling, was resolved by the microscope into innumerable crystals of oxalate of lime.*

These observations have been repeated in other cases of excess of uric acid, and with like approximate results. Thus, in the space of four days, in the presence of an alkaline salt, the large excess of uric acid disappeared and was replaced by an equally large proportion of urea and oxalate of lime. We thus explain the mode by which excess of uric acid may be converted into a soluble and easily excreted element of the urine, and an explanation is offered of the cause of the defective elimination of uric acid in gout and renal calculus, parallel to that offered by Miahe for the presence of sugar in the urine of diabetic patients, that the blood being deficient in alkaline salts, the presence of which is necessary for the oxidation of the sugar derived from the amylaceous elements of food, no further conversions of the sugar into oxalic acid, and still further into carbonic acid and water takes place, but as sugar, it appears in the urine.

These facts in relation to uric acid may be thus summed up. Physiologically it has been observed that uric acid injected into the stomach or blood-vessels increases the amount of urea and oxalic acid in the urine. Chemically it has been shown that uric acid, by a process of oxidation in the presence of an alkali, is converted also into urea and

* See further the subject of Oxaluria.

oxalic acid. Hence, it may be concluded that the accumulation of uric acid in gout and calculus arises from a defective process which fails to carry forward the conversion of uric acid into a soluble form for excretion, while, clinically, it has been observed, that in cases of excess of uric acid, that if an alkali be supplied and the patient placed under circumstances favorable for the increase of the oxidation of the tissues, the excess of uric acid, and the ill consequences of its accumulation rapidly disappear. This subject will be further discussed under the head of oxaluria.

Explain the modus operandi how we may it is an indisputable fact that potash salts do immensely facilitate the reduction of uric acid and the conversion of it into a form capable of excretion by the kidneys, probably in the shape of an increase of urea. I therefore unhesitatingly offer these facts as leading to a correct interpretation of why it is that these alkaline waters are so undeniably efficacious in the treatment of gout and calculous disorders of the kidneys; and why this efficiency is so much greater when drunk at their source than when taken either as exported from the springs, or as factitious imitations of them, however chemically correct they may be in their composition. The early hours, the large period of the day spent in the open air, the usually elevated position of these springs, the simple diet and cheerful amusements provided for the visitors, all play their part doubtless, but the oxidizing agency of the pure clear air with proportionate exercise, in conjunction with the alkaline material circulating through the organism are the real operating causes in reducing the uric acid, or converting it into a form more easily excreted.

§ 14. PYELO-NEPHRITIS and PYO-NEPHROSIS, *from retention of urine, from stricture or disease of the prostate or bladder.*— *Symptoms.*—This form of renal disorder, caused by obstruction to the flow of urine from the bladder, either by stricture of the urethra, or disease of the prostate or bladder, is of slow and insidious development, and, for the most part, requires that the obstructive cause should be of long standing, and the patient long the subject of difficult micturition; sufficient, indeed, to necessitate the frequent use of instruments for the withdrawal of the retained urine. The diagnosis of these cases rests chiefly on this fact—the pre-existence of difficult micturition for months, perhaps years. The cases are rare in which pyelo-nephritic symptoms are developed after a single or the first attack of retention of urine. The case recorded by Mr. Stanley is a rare one. Usually the sufferer from permanent stricture, who, for a long period, has had occasional attacks of retention of urine, and which have been relieved only by surgical aid, notices that his urine, when drawn off or passed without an instrument, is cloudy and purulent; it contains a large amount of pus and mucus, mingled with shreddy membranous matter. It rapidly becomes ammoniacal and ropy. By the microscope, pus cells, exudation or mucous cells, modified epithelial cells from the pelvis and ureters, with crystals of the triple phosphates are seen. In the event of the mischief having reached the cavity of the kidney, and from the state of the lower portion of the ureter preventing a free escape of the urino-purulent fluid, a renal tumour may be formed and be detected; and conditions such as are observed in calculous nephritis may be present. The diagnosis will rest on the previous history of the patient. The pre-existence of stricture and of attacks of retention of urine favouring the probability of pyelitis from this cause rather than from calculus. The antecedents of the patient throw much light on the diagnosis in both cases. Calculous nephritis has its well-marked precursory symptoms, which can never be mistaken for those present in these cases of pyelo-nephritis from stricture, &c.

The treatment of these cases is chiefly surgical. To keep the urethral channel free and open is obviously the first point. To reduce the purulent or muco-purulent flux from the pelvis of the kidney, the ureters and bladder is the second. To maintain or restore the general health is a primary object, for this disease occurs mostly in constitutions broken down by irregular and dissipated habits of life, and often aggravated by the influence of tropical climates. The worst cases I have met with have been officers or men from the East or West Indies. Until the constitution has gained some vigour, it is idle to expect any diminution of the renal symptoms. Premising, however, that the chances of retention have been reduced by surgical treatment to a minimum, a generous diet should be ordered, stimulants allowed in moderation, but only those which are aids to nutrition. Quinine and iron may effectually contribute to the effects of a nutritious diet. As the general health of the patient improves, he may try those remedies which have had some reputation in controlling these muco-purulent fluxes of the urinary apparatus. The infusion of the Pareira Brava, or one or two ounces of the infusion of Buchu, may be taken three or four times a day. The prospect of recovery must rest entirely on the probable extent in the urinary apparatus to which the irritation has reached, and consequently will bear a direct relation to the period of time over which these urethral and urinary disturbances have extended. The effect of long standing prostatic disease is not less capable of setting up vesical irritation, extending eventually to the ureters and kidneys. Such cases are generally met with amongst those advanced in life. Sometimes they are complicated with renal calculus, and the diagnosis becomes the more difficult. The mystery of these complications can only be cleared away by a correct account of the patient's health for the whole period he may have suffered from urinary disorder. If the patient has been previously subject to calculous disease of the kidney, or has suffered from gravel or gout, and if the urine has been bloody or purulent before the prostatic symptoms distressed him, the presumption will then be that the renal disorder is due rather to the pre-existent disease than solely to irritation transmitted from the diseased prostate. These cases admit only of palliative remedies.

§ 15. TUBERCULAR NEPHRITIS—TUBERCULAR PYELITIS— *Symptoms, Diagnosis, and Treatment.*—There are no symptoms, in the absence of a renal tumour, by which, during life, nephritis of a tubercular origin can be distinguished from pyelitis from a similar cause. It may, indeed, be affirmed that the tubercular matter most frequently has its origin in the cortical portion of the kidneys and extends thence to the lower infundibulum and ureters, so that such case of tubercular pyelitis, in its origin, may be one of tubercular nephritis. It is only when an enlargement of the kidney becomes apparent, when a tumour in the position of one or other of these organs is detected, accompanied by purulent urine, that a distinction can be made between nephritis and pyelitis from a tubercular cause.

The diagnosis between tubercular pyelitis and calculous pyelitis is of more pathological importance.

Practically there is more difficulty in this diagnosis than might be supposed. Indeed, it rests on such slight distinctions that the most experienced in these renal disorders may be easily deceived. It is only from the antecedent symptoms, dating from the commencement of any urinary trouble, and an accurate record of their progress and succession, that a trustworthy diagnosis can be formed. In both diseases there is purulent urine, and there may be a renal tumour. Nor does the purulent urine differ in any of its microscopical characters one from the other. The presence or absence of a certain granular matter supposed to be characteristic of tubercular disease is not trustworthy. Crystals of double and triple phosphate are commonly present in both, and depend on molecular changes in the chemical composition of purulent urine which may have taken place in the bladder or at certain seasons of the year, after being passed through the agency of temperature alone. No microscopic test can establish a distinction. The observer is, therefore, thrown back on the character and sequence of the antecedent symptoms. If reference be made to the subject of calculous nephritis, it will be seen that, in the great majority of cases, the urine exhibits, through a succession of stages, various characters. Scanty urine, more or less troubled with sand or gravel, or with slight traces of blood, frequently passes, and as its quantity becomes increased and its specific gravity lower, a mucous cloud becomes more or less apparent, presenting at this stage examples of epithelial cells from the ureters and infundibulum, the urine soon becoming milky even when passed, and alternately resolving itself into a purulent urine with a well-defined sediment of pus cells. These phases of the symptoms of urinary disturbance are slow and progressive, passing from one to the other in the sequence described, and covering periods not of weeks but of months, and even longer, before the full development of purulent urine and a renal tumour. On the other hand, in tubercular disease of the kidney, no such protracted interval is observed between the first indications of renal disturbance and the voiding of a urine loaded with pus. The urine may be said almost immediately to become purulent. The following may be accepted as the characteristic symptoms in a typical case of tubercular nephritis. The general health is observed to fail, a sense of weight or uneasiness about the loins is experienced, micturition becomes more than usually even painfully frequent, and an attack of hæmaturia may follow. When this happens a rigor, more or less marked, attends the appearance of blood in the urine. In some cases nausea and vomiting may be present at the earliest stage. Hæmaturia, however, is not always present, for this hæmorrhage bears to renal tubercle the same relation which hæmoptysis does to pulmonary tubercle. Though a frequent it is not a necessary symptom. The early stage of renal tubercle is marked by a degree of suffering far in excess of what is witnessed in the early stage of calculous nephritis. A most distressing pain is felt at the neck of the bladder and extremity of the urethra, with a sense of weight and dragging in the perinæum, accompanied by a most urgent need to pass water. A small quantity is passed with a momentary and transient relief, to return again in a few minutes, with the same feelings of pain and distress and the sympathetic urging to micturition which may recur at longer or shorter intervals, every few minutes to half an hour, throughout both day and night. These symptoms are far more aggravated, more intense, and the suffering of the patient more plainly expressed in the countenance, than is observed in calculous disease of these organs. Moreover, the urgent frequency of the desire to pass water, with scarcely an interval of ease or freedom from agonising, crushing pain, an irritable stomach, with nausea, or, at least, inappetency, sleep entirely banished, and scarcely obtained by the agency of opiates, soon tell on the nutritive condition of the patient, and loss of flesh and rapid emaciation plainly declare a contrast between this form of renal disease and that derived from calculous deposit. It is a singular clinical fact that in calculous disease of the kidneys there is scarcely any appreciable departure from the ordinary aspect or condition of the patient, and except during the passage of a calculus down the ureter, or the displacement of one within the kidney when the stomach becomes irritable and vomiting may occur, but there is little or no disturbance in the digestive organs, nor any diminution of the average nutrition of the body. It is far otherwise in tubercular disease. Wasting and emaciation are apparent from the beginning, as in other forms of tubercular disease running an acute course, the pulse increases in frequency and maintains throughout the disorder a rapidity equal to that in phthisis pulmonalis. In the great majority of cases a careful examination of the lungs fails to detect any indication of latent tubercle. The sounds of the heart are natural, and usually there are no symptoms of disease in other organs than the kidneys. The urine at the commencement of the tubercular activity may contain blood, or it may show mucus in considerable excess of what is present in health. The presence of blood is apparently caused by the inflammatory engorgement of those parts in which the tubercle is passing into a stage of softening and purulent change, and the rupture of some of the glomeruli and the passage thence of blood into the urinary tubes, and thence into the bladder. Microscopically there is nothing to distinguish this hæmorrhage from others having an origin in the kidney. Nor does the excess of ordinary mucus in the urine exhibit through the microscope any characters to distinguish it from that which is present in the early stage of calculous nephritis. The objects consist of the varieties of epithelial cell peculiar to the urinary passages. Hence a careful examination will detect every form, from those peculiar to the urethra and bladder, squamous or pavement epithelial cells, to those lining the ureters, pelvis, and calyces of the kidneys, the latter pyriform or even spherical, and half the size of the former. No object of diagnostic value is seen, for these are common to the early stage of calculous disease, and are expressive only of a diffuse sympathetic irritation throughout the whole urinary apparatus, excited by the commencement of active suppurative disease in the kidney. It is now, however, within at most a few days of the local manifestations of disturbance expressed by the lumbar, pubic, perineal, and urethral uneasiness, the frequency of micturition and the blood-stained or mucous urine, that the almost sudden appearance of a large proportion of pus in the urine offers the suspicion that the rapid development of pus can only arise from tubercular abscess of the kidney. No other disorder capable of producing purulent urine acts with such rapidity or with symptoms following each other in such order. Night sweats and evening exacerbations of febrile action may excite attention and aid materially in forming a correct diagnosis. It is thus seen how necessary to an accurate diagnosis is the early history and the sequence of the symptoms. The most characteristic is the rapid appearance of urine highly charged with pus, within a very short period of the first indications of renal trouble. The next stage of the disease is the development of a renal tumour; from the commencement, if one kidney is alone affected, the patient experiences considerable uneasiness on pressure being made in the lumbar space. A great increase of pain results, and a pungent darting anguish on each act of pressure in the direction of the ureter and to the groin or pubis; a sense of nausea is often created by this examination. The sympathetic conditions of the cutaneous nerves of the lower limbs are not so well marked in tubercle of the kidney as in calculus. There are seldom those feelings of numbness or of altered sensation in the surface supplied by the external crural cutaneous nerve. Pain and aching is sometimes referred to the anterior crest of the ilium, but this is for the most part obscured by the greater pain in the neck of the bladder, urethra, and perinæum, and the wearing frequency of micturition. In a short time, a few weeks and even less from the earliest symptoms, a slight fluctuation may be perceived in the lumbar space, and corresponding degree of fulness in the abdominal walls. A careful examination will now detect a tumour occupying the region of one or other kidney, any pressure on which gives rise to an aggravation of all those sympathetic pains which have been enumerated.

Severe rigors of daily recurrence, great irritability of the stomach, a constant nausea and obstinate vomiting after the simplest ingesta, and increasing emaciation and exhaustion, drenching sweats, tormenting thirst, with undiminished irritability of the urinary passages, mark the closing stage of these melancholy cases. The pus may attempt to burrow and find an outlet if the occluded ureter prevents the free discharge of pus, or the tumour increasing in size from a like cause acts on the intricate plexus of nerves whose ganglia exist in such number in and around the region of the kidneys.

Convulsions may come on, tetanic almost in character, consciousness remaining intact throughout the most terrible of such paroxysms. In other cases so painful a complication is spared the unhappy sufferer, but increasing exhaustion, a failing pulse, a death-like pallor, and scarcely perceptible breathings, mark the easy and painless transition from life to death.

Prognosis.—The progress of tubercular disease of the kidney is so little under the influence of therapeutic means that any prospect of recovery, or even amelioration of the disease, is very remote. It need hardly be said, therefore, that the prognosis in every such case is essentially unfavorable.

Treatment.—The treatment of tubercular nephritis and pyelitis should chiefly embrace two objects:

I. The relief and mitigation of the pain and distressing irritation of the urinary organs.

II. Through the agency of nutritive and restorative means to extinguish, if it be possible, the tubercular crasis, and thus arrest the development of fresh morbific material out of which the destructive and suppurative process issues.

The various preparations of opium are the only remedies on which reliance can be placed to obtain any palliation to the sufferings of the patient in this distressing disease. Other remedies are all but powerless. It signifies very little which preparation of opium is selected. The Liquor Opii Sedativus, and one or other of the liquid preparations of morphia, the Liquor Morphiæ Acetatis, or the Liquor Morphiæ Hydrochloratis, are the most convenient. It is desirable to give the narcotic in the fluid form, as it operates more quickly. The dose will, of course, depend on the age and susceptibility of the patient to the influence of the drug. At first, moderate doses, eight to ten minims of the Liquor Opii, or fifteen of the preparations of morphia, may be given at intervals till some diminution of the pain and urgent calls to micturition are apparent. Larger doses will, as the disease advances, be required and easily borne. The use of diluents, so efficacious in calculous disease of the kidneys, is of but little avail in this disorder. The infusions of Pareira, of Buchu, the decoction of *Uvæ ursi*, that of *Chimaphilæ umbellatæ*, are equally inert, nor, if the pathology of the disease be kept in view, could it reasonably be expected otherwise. It is very evident that remedies acting only through the urine, by imparting to it such astringent or volatile principles as may be contained in them, can be of little service, and could neither arrest the deposit of fresh tubercle nor check the suppurative process which follows. It is only on an improved quality of the blood and the nutritive processes sustained by it that any curative expectation can rest. Consequently to that class of remedies known as hæmatics, in conjunction with such food as the digestive organs of the patient will bear, we must almost exclusively look for power to check the advance of this formidable disease.

It has been observed in the section of the pathology of this disease that obsolescence of tubercle-pus takes place only on the arrest or extinction of the tubercular dyscrasia. The hope of establishing this crisis in the progress of tubercular disease of the kidneys must rest on those general principles of treatment which are efficacious in phthisis pulmonalis. The influence of climate, however, which plays so important a part in the pulmonary form of disease may be considered of minor importance in tubercle of the kidney. A careful attention to the clothing of the surface of the body with flannel, a cautious restraint to the house, and the uniform temperature of a well-warmed apartment in the winter time, are measures which should be strictly enforced. In the event of any diminution of the daily amount of pus excreted with the urine, or any mitigation of the local pain, or subsidence of the frequency of micturition, these cautious restraints should be enforced with increased vigilance. Presuming on the mitigation of the more urgent symptoms and the subsidence of the more painful ones, patients neglect precautions which are as necessary, perhaps even more, in the decline of the suppurative process as in the very height of its activity. Almost every case of tubercular disease of the kidney presents these points of mitigation, and although it be a very rare exception to meet with one which eventually recovers, yet the writer has within his experience seen two where it may fairly be affirmed that the rigid perseverance in every precaution necessary in the more active stage of the disease did most essentially contribute to an ultimate disappearance of the pus, and eventually to apparent recovery. On the other hand, examples of the opposite character are not wanting. A period of the most promising mitigation of symptoms has occurred, the amount of pus has become noticeably less, the weight of the body has plainly increased; but this gleam of hope has brought with it too early a confidence in the subsidence of the disease; a return to the active duties or the pleasures of life has prematurely occurred, and within a short period a renewal of all the more serious symptoms has blighted the promise of recovery, and carried the patient with increased symptoms of aggravation to the grave.

As in phthisis pulmonalis so in phthisis renalis, the digestive organs, both as regards appetite and the assimilative function, fail, partly due to the sympathy existing between these organs and the stomach, but chiefly to that depressed state of the gastric function, the prominent characteristic of tuberculosis, to which the general name of strumous dyspepsia has been applied. This state of defective digestion prevails in the tubercular disease of the kidneys in a marked manner, and it forms one of the greatest obstacles to the influence of diet in moderating the tubercular or scrofulous diathesis. It must be very clear that the power of iron in any of its forms to restore to the blood any of the more important characters of which in tuberculosis it is deficient must depend on the co-operative influence of food. If food cannot be taken chyle cannot be formed, and if chyle is deficient iron is useless. Nevertheless it is advisable in these cases to give steel in some form or other, however small the quantity of food may be.

In those cases or at that stage of the disease where vomiting and the rejection of ingesta of all kinds takes place ferruginous remedies must be discontinued, and resumed again on the stomach regaining its power to take food; still, the amount of iron given should, in great measure, be proportional to the amount of food taken—small doses, if the proportion of food be small, larger as the powers of digestion increase.

The kind of food selected in those cases is of importance. It is occasionally observed in the dyspepsia of tuberculosis that there is a positive abhorrence to fatty food. This is so important an element in the nutrition of those who are the subject of tubercle that all means must be tried to introduce it into the system.

It is with this view that cod-liver oil is employed. It is best to begin with small doses, a teaspoonful, increasing gradually to two or a table-spoonful, but larger doses in these cases of tubercle of the kidney are not often tolerated. It is well to administer it in conjunction with some form of iron. It may be given with doses of the citrate of quinine and iron, or the citrate of iron may be given alone with the oil. These preparations of iron may be considered as the best vehicle for the administration of the oil. Should at any time during the progress of the disease the patient be able to take animal food, and it often happens during some period of mitigation that the patient can do so, iron in some other form is better. A factitious chalybeate water may be extemporised, and is often grateful to the patient, by adding from five to ten drops of the tincture of the sesquichloride of iron to half a tumbler of seltzer water, and drinking it before the water has ceased to effervesce. This should be taken from half an hour to an hour after food.

§ 16. CANCEROUS NEPHRITIS.—*Diagnosis and symptoms.*— Cancer may be deposited in the renal structures. Its development is usually marked by frequently recurring attacks of hæmaturia, without any of the typical symptoms of calculus or tubercle.*

As the disease advances, and the deposit softens and breaks down, if the variety be encephaloid, cancer-cells may be seen in the urinary sediment. In scirrhus or hard cancer the diagnosis is more difficult, but this is of rare occurrence. In encephaloid the urine may become loaded with a purulent deposit. The ureters may be unable freely to discharge the cancero-purulent material, and a renal tumour be ultimately formed, such as occurs both in calculous and tubercular disease of these organs.

The diagnosis will greatly depend on the detection of cancer-cells in the sanguinolent or purulent sediment.

§ 17. NEPHRITIS—*Endemic; endemic hæmaturia; parasitic.*—*Symptoms and treatment.*—This form of hæmaturia was formerly thought to be derived from the bladder alone. Dr.

---

* See a case, vol. x, 'Pathological Trans.,' p. 188.

Todd* believed the source of the hæmorrhage to be from the bladder, and that it seemed to be essentially a catarrh of that organ, with occasional hæmorrhage. The observations of Dr. J. Harley, as already mentioned, have conclusively shown that the kidneys are invaded as well as the bladder, and that the hæmorrhage, in all probability, is chiefly derived from them. The most reliable diagnostic symptom is the period during micturition in which the blood makes its appearance. The first portion of urine passed is free from blood. It is not even discoloured. It is with the last drops of urine that the blood makes its appearance, a few drops of pure blood passing with the action of the ejaculator muscles.

The following are the most characteristic symptoms recorded by Dr. J. Harley.† The first indication of the disease is the passage of a little blood after emptying the bladder of clear urine, usually not more than a few drops, and never exceeding a teaspoonful. In the case under observation the hæmorrhage occurred after every act of micturition for the first fortnight, and subsequently it continued for nearly four years, with weekly or fortnightly intermissions. Exercise caused a slight increase. Small filamentous bodies, irregularly cylindrical and blood-stained, occasionally colourless, made their appearance in the urine in the last year. There was an increased frequency of micturition, but no irritability of the bladder, and the quantity of urine passed was the ordinary average. The complaint appeared to occasion little annoyance, and the general health was unaffected.

The filamentous bodies, when microscopically examined, were found by Dr. Harley to contain the ova of the nematoid parasite, the Bilharzia.

Dr. Harley's suggestions with regard to treatment are valuable. He thinks that remedies which pass out of the body unchanged in the urine are those on which reliance should be chiefly placed. Belladonna and hyoscyamus are eliminated unchanged by the kidneys. In their passage through these

* 'Urinary Diseases.'
† 'Med.-Chir. Trans.,' vol. xlvii, p. 57.

organs they may render the ova abortive and incapable of further development.

The most serious result of the infection of the urinary organs by this parasite is the formation of renal calculus, which has been so frequently observed in Egypt, and examples of which are not wanting in those from the Cape, as well as from the Mauritius.

§ 18. PERI-NEPHRITIS.—*Symptoms, diagnosis, and treatment.*—When the textures surrounding the kidneys partake of the effects of external injury which has torn, crushed, or injured these organs, the symptoms, in addition to those already enumerated under the head of nephritis from external violence, will be tumefactions and ecchymoses about the lumbar region, with considerable tenderness on pressure or any movement of the body; beyond these and the evidence of severe shock there will be nothing typical. The symptoms will necessarily be regulated by the extent and seat of the local injury; and as no two cases in these respects can be alike, each case will require careful examination before an opinion can be with certainty offered as to the area of mischief which the accident has caused.

Peri-nephritis from renal abscess is by no means infrequent; it mostly occurs as a sequel to calculous or tubercular pyelitis, particularly from the first; and, as already stated under the head of causes, arises from the ureter becoming occluded, and thus preventing the escape of the urino-purulent contents of the kidney from passing through the bladder. The symptoms of this disorder will be those of calculous or tubercular pyelitis as the forerunners, added to a sudden diminution or total disappearance of pus from the urine, corresponding, in point of time, with rigors, uneasiness, pain, a sense of fulness, even of dragging weight, in the region of the kidney affected, and the gradual rise of a distinct enlargement or tumefaction in the lumbar or adjacent region. With the slow increase of the size of the swelling a sense of fluctuation is detected.

The direction taken by the purulent fluid varies in different cases. In some the swelling may be somewhat circumscribed, but prominent, situated to the right or left of the spine in the lumbar region; or it may be immediately over or in contact with the brim of the pelvis posteriorly; it may be slow in coming near the surface, and it may burrow between the fascia of the gluteal muscles, and be felt in the buttock laterally, even as low down as within an inch or two of the head of the hip bone; or there may be no external manifestation of a tumour. The purulent urine has suddenly ceased; for a day or two no particular symptom occurs; the patient then suffers severe rigors, with general distress, inappetency, restlessness, quick pulse, and some fever. No tumour can be detected externally, but careful exploration of the abdomen will reveal a swelling, enlargement of one of the kidneys, oftentimes very considerable, and not difficult of detection. Coupled with the sudden disappearance of pus from the urine, the origin of this tumour is easily accounted for, and so far the symptoms of fever and restlessness explained. Peri-nephritis exists, but in what direction the pus will escape no experience can predict. In one case under my observation, with all these symptoms, a sudden diarrhœa came on, which effectually emptied the renal tumour, and declared unequivocally that the abscess had found its way into the colon by an ulcerative process, and the pus escaped by the bowels. Other cases, immediately fatal, have followed the escape of the contents of the renal cyst into the cavity of the peritoneum. Such are the various directions, and even others have been observed, in which a renal abscess may find an outlet. It is more, however, to the purpose of this section to speak of the treatment of such cases as permit of probable relief. A case which recently came under my observation will sufficiently illustrate what should be done, and the amount of relief to be obtained in similar instances. A gentleman, about forty-five years of age, had been ailing for some months. He had suffered from frequency of micturition, pain both at the extremity of the meatus as well as at the neck of the bladder, with urine stated to be distinctly purulent. Suddenly the urine became clear, depositing, on cooling, abundance of lithates. This change in the quality of the urine was quickly followed by some constitutional disturbances, irritability of the stomach, nausea, and inappetency, and a sense of fulness, of a dragging weight in the right lumbar region. When I was asked to see him four days had passed since the rigors and the first perception of the swelling in the lumbar space. I found the symmetry of the loins altered by a visible swelling on the right of the spine, filling all the dorsal space, extending laterally to the brim of the pelvis, overlapping the brim and reaching downwards almost over the glutæus to the trochanter. There was a distinct sense of fluctuation everywhere in this diffused swelling; and pain of an acute kind followed pressure anywhere, but more distinctly in the most prominent part of the swelling. The patient could not stand upright from the painful downward drag this position gave to the right kidney apparently. The diagnosis of this case was not difficult, for it was very instructive to make out that the renal fluctuating tumour, with rigors and feverishness, corresponded remarkably with the all but sudden disappearance of the pus from the urine. An exploratory needle was inserted into the most prominent part of the swelling, and pus being found in the groove, a free opening was made into the swelling, and a quart or more of a urino-purulent fluid escaped. Opiates were given for a few nights. The patient quickly recovered his appetite, and, after a sojourn of some weeks at the sea-side, was able to return to his ordinary occupation. I am told the urine is now perfectly natural, and that he suffers no symptoms of his former malady. His system is working with one kidney; the right one, from its probable sacculated condition, having collapsed or contracted, and shrivelled up without a vestige of tubular structure being left. Examples of this shrivelling or atrophy of the tubular and cortical part of these sacculated kidneys may be met with in almost every pathological collection.

Rokitansky speaks of peri-nephritis as sometimes arising from the extension of a nephritis to the fibrous capsule of the kidney, and involving the adipose layer in which the kidney is imbedded. He also states that it results from wounds or any external injury affecting the kidneys, as well as from urinous infiltration, as has been illustrated by the preceding case. The symptoms of these cases are sometimes doubtful and obscure, except in cases of external violence.

§ 20. NEPHRITIS—*Induced by, or associated with, pregnancy.—Symptoms and diagnosis.*—It has been already remarked under the section of causes that the pressure of the gravid uterus may induce such a disturbance in the circulation through the kidneys as to bring about, not only œdema of the lower extremities, but even to render the urine albuminous. When this latter symptom occurs in a pregnant woman it is a subject of great anxiety to the experienced obstetrician, for he knows full well that this may be a symptom of one form of morbus Brightii, and, if so, that the period of delivery, if abortion does not previously occur, is full of hazard and danger to his patient, from the all but certain occurrence of puerpural convulsions, with probable fatal results. Can a safe diagnosis be made in a pregnant woman between albuminous urine arising simply and only from mechanical impediment to the circulation through the kidneys, and that which, while it arises partly from this cause, is also aggravated by antecedent renal disorder, although the patient has not, previous to pregnancy, exhibited any symptoms, or at least none have been noticed, of this proclivity to renal disease?

The diagnosis must rest exclusively on the microscopic character of the sediment in the urine. In cases of albuminous urine from obstructed circulation through pressure only, the sediment, when examined by the microscope, will consist of a few simple granular casts, unaccompanied by evidence of cell-structure derived from the tubes. Some vaginal or urethral epithelium is usually seen. But if the albuminous urine be derived from kidneys which are the seat of special pathological changes apart from, and independent of, the effects of pregnancy, then the sediment will testify to these morbid processes; and casts, with more or less of free fat-granules, with large spherical corpuscles loaded with fatty and granular matter, as well as renal epithelial cells, either free or contained apparently in the casts, will be readily found.

I believe this microscopic examination of the sediment to be the only means for a correct diagnosis between an albuminuria in pregnancy simply congestive and the same state of the urine arising from serious, and all but fatal pre-existent disease.

§ 21. Pyo-nephrosis.—The subject of purulent abscess, or a sacculated condition of the kidney from retained urino-purulent fluid, has been treated in the section on Peri-nephritis.

Hydro-nephrosis most usually is a congenital condition arising from malformation of the outlets or passages from the kidney. Such cases have been found, however, in adults. See a case reported by Dr. Roberts.*

* 'Urinary Diseases,' p. 408.

PART II.

CHRONIC NEPHRITIS—CHRONIC ALBUMINURIA.

GROUP II. *Renal diseases non-inflammatory*—without primary symptoms of inflammation.—Urine albuminous, rarely purulent. Sediment containing specific microscopic objects—diagnostic casts of the uriniferous tubes. Diseases chiefly characterised by evidence of degeneration of the glandular, the vascular, or fibrinous structures of the kidneys—granular, fatty, fibrinous, waxy, or amyloid. Some of this group associated with, or even arising from passive engorgement of the kidneys from obstructed circulation through the heart, lungs, or liver; more or less of dropsy accompanying them.

CHAPTER IV.

Chronic Albuminuria.

The following four varieties of post-mortem conditions represent the structural changes of the kidneys in the group of diseases designated chronic albuminuria or chronic morbus Brightii.

I. The small red, contracted, granular kidney. The cirrhotic kidney of Dr. T. Grainger Stewart and Dr. Harley.

II. The large granular fatty kidney.

III. The amyloid kidney.

IV. The atrophic, contracted, nodular, gouty kidney.

Before proceeding to describe in detail the anatomical characters of these varieties, I deem it desirable that the student should understand the ground on which this arrangement is based, and the reason why a distinction is made between the red contracted granular kidney and the atrophic contracted nodular kidney, which are by many writers recognised as one and the same. The first variety is sometimes called the cirrhotic, and sometimes the gouty kidney. It is, doubtless, frequently met with in patients who may have had gout; but it is as frequently, perhaps more frequently, seen in those who have not. When it occurs associated with gout it is also connected usually with lead impregnation, it being the usual post-mortem condition of those who have suffered both from lead poison and gout. But the atrophic nodular contracted kidney is never seen but in those who have not only had gout, but who have suffered conspicuously from the disease in the shape of chalk-stones, or tophaceous deposits in the fibrinous and cartilaginous structures. In these kidneys the deposit of urate of soda is present either in the tubes, the interstitial fibrinous structure, or in the coats of the arteries as opaque atheromatous spots.

The large granular fatty kidney, the second variety, differs from the somewhat similar post-mortem condition traced to inflammatory action, in that the disease has not an active inflammatory origin, but is traceable to obstructive conditions in the circulation through the kidneys from pulmonary or cardiac diseases, with probable accessions of congestion during its progress.

The amyloid variety is sufficiently distinct in its characteristics, and is now admitted into every classification of renal disorders. Dr. Wilks, Dr. Grainger Stewart, and Dr. Dickinson, have severally contributed greatly to the elucidation of this form of renal disease.

Chronic Nephritis.—In this group of renal diseases the term chronic nephritis may seem inappropriate, as the word implies an inflammatory process. It, perhaps, would be more consistent with modern views to substitute the more general term chronic albuminuria, or to accept as synonyms chronic morbus Brightii, or non-inflammatory albuminuria, or chronic renal degeneration.

Chronic albuminuria is the most intelligible, as it simply implies a quality of urine which is present in every form of renal disease belonging to the present group.

Chronic Albuminuria. — *Causes and Pathology.* — The following may be enumerated as the chief predisposing causes of chronic renal disease.

Pre-existing blood poison.—1. Certain fevers. 2. The strumous taint. 3. The syphilitic taint. 4. The gouty taint. 5. Certain mineral poisons, lead and phosphorus. 6. Intemperance through alcoholized blood. 7. Obstructive conditions in the circulation of the blood, occurring in diseases of the heart and lungs, and occasionally of the liver. The loss of alkaline salts from the blood by purulent drain—Amyloid—Depurative disease of Dickinson.

CHRONIC NEPHRITIS.—Synonyms.—*Chronic albuminuria; chronic Bright's disease; chronic desquamative nephritis; non-inflammatory albuminuria; chronic granular degeneration.*

The line of demarcation between acute and chronic disease of the kidney is arbitrary rather than real. Every acute form of disease may terminate either in recovery, death, or in change of structure, so protracted as in the view of many observers to become classed under the term chronic in contradistinction to the acute or more rapidly advancing diseases. It will be found in the preceding part of this work, that forms of disease of the kidney, in the class of Bright's disease, have been classified as acute, which, according to other authors, are designated as chronic; this apparent discrepancy is thus explained. Those forms of Bright's disease which have their origin in an inflammatory or active state of congestion, and which run their course with varying degrees of intensity and rapidity, are classified as acute, quite irrespective of the period of time through which they advance to a fatal termination. Thus the form in which the kidneys become large, pale, and anæmic, commences with symptoms of inflammatory engorgement; but may run a course of many months before arriving at a fatal termination. There is a continuous morbid process from the first. A disease so protracted in its course in the sense of duration, is certainly chronic. But as its origin is in an acute process, and the progress however slow, continuous and not intermittent, and as its products are inflammatory, it is only consistent with an effort to arrange these varieties of renal disease in some systematic order that they are placed in the catalogue of the acute forms. The difficulties of such an effort of classification into acute and chronic varieties are increased by the frequently recurring interference of congestion, or inflammatory action supervening on the more strictly defined chronic forms; thus presenting results partly due to an insidious chronic process, and partly to inflammatory action grafted as it were on it. This is exemplified in the large mottled granular kidney often associated with cardiac and pulmonary dropsy, where the fibroid element and the structural changes in the walls of the blood-vessels belong to an antecedent insidious chronic process; while the increased volume of the gland, and the granular character of the deposit, similar to what is seen in the acute form of morbus Brightii, connect it with the inflammatory exudation of that variety.

The commencement of the chronic forms of albuminuria is insidious, and often impossible to trace. The transition from an average state of health, or from a state in which no marked symptoms existed, to one in which disturbance in the urinary secretion is recognised, is unmarked by any special or characteristic group of symptoms.

Headache, loss of physical energy, wandering, pseudo-rheumatic pains, disrelish for food may testify to failing health, but do not pointedly imply renal disorder. Unlike, therefore, the acute form, which takes its rise in feverishness and scanty and bloody urine, signs unmistakeably emanating from the kidneys; the chronic forms may have been in silent development months before they are recognised. It is on these facts that I would base the distinction between acute and chronic albuminuria. This group of chronic diseases includes the most important varieties of those diseases of the kidney which are named after the distinguished physician who first gave them a place in pathology. There have been many workers in the field since Dr. Bright's original observations. The result of their united labours has been to gather together a vast mass of facts, upon which the pathology of this disease now rests.

To enumerate all those who have contributed to this result is beyond the scope of the present work. It would include the names of the most distinguished French, German, and English pathologists. Nevertheless, I feel it an act of duty, and for the advantage of the student in his clinical studies, to mention those English writers and pathologists from whose observations myself and others have derived so much information; and if from some I occasionally differ, it is the more agreeable to acknowledge the great amount of pleasure and instruction that I have derived from their writings.

From Scotland came first, and not the least distinguished, of those who followed directly in the wake of Dr. Bright's original observations.

Dr. Christison, in his work on 'Granular Degeneration of the Kidneys,' contributed greatly to the diffusion and extension, both at home and abroad, of a knowledge of this disease. And it is worthy of the eminence of the Scotch school of pathology, that only last year fresh illustrations of the nature and treatment of Bright's diseases should have been offered by Dr. Grainger Stewart, whose researches into the nature of amyloid degeneration of the kidneys, places him in the very first rank of those who have removed much that was obscure in the pathology of what was formerly styled the waxy kidney.

The teachings of Dr. Bright in the very home of his investigations in Guy's Hospital, did not fall on unwilling or inattentive ears. The contributions of Dr. Owen Rees, not only on the subject of albuminuria, but on other forms of renal disorder, gave to the subject of renal pathology an impulse which the writings of Dr. Prout and Dr. Golding Bird had to some extent started. With Dr. Owen Rees, however, rests the unquestionable merit of disallowing several theoretical errors into which both those writers had fallen, and which recent investigations have fully endorsed.

Among the names of English pathologists who have enlarged our knowledge of this group of diseases, a place of pre-eminence belongs to Dr. Geo. Johnson. Throughout this work I have had frequent occasion to refer to his researches, the most brilliant of which will be found scattered through the 'Transactions of the Royal Medical and Chirurgical Society,' during the last twenty years, as well as many valuable reports on cases in the 'Transactions of the Pathological Society.'

Among the more recent writers on the subject of Bright's disease, and whose researches have added many points of instruction to our former store of facts, may be mentioned Dr.

Wilks, of Guy's Hospital, Dr. Goodfellow, Dr. George Harley, and Dr. L. Beale, and especially Dr. Roberts of Manchester.

Dr. Dickinson, of St. George's Hospital, availing himself of the valuable field of observation which his position, first as medical registrar, and subsequently as assistant-physician, afforded him, has, in his valuable work 'The Pathology and Treatment of Albuminuria,' added largely to our knowledge of the structural changes in these chronic forms of morbus Brightii.

The illustrations in this work are admirably executed and most instructive. His classification of these diseases is simple and certainly very intelligible, and clears up a great deal of the difficulty to which an arbitrary division into acute and chronic forms often leads. The basis of this arrangement rests on the assumption that all forms of albuminuria may be referred to one of three elemental sources of disorder. First, either in the renal tubes and their epithelial or gland cells; or, secondly, in the fibrous structure or tissue which holds up, and keeps in position, Malpighian bodies, blood-vessels, and uriniferous tubes; and, thirdly, in the blood-vessels, the minute arteries being the starting-point from which the infiltration of the whole organ with amyloid substance takes place.

The renal tubes, the fibrous tissue, the blood-vessels, are, therefore, the three pathological centres of disorder, and to one or other of these each form of disease may be referred. The disease, if inflammatory, is tubal nephritis. The morbid change residing exclusively in the renal tubes, convoluted or straight, and their epithelial cells, and the disordered action (the inflammatory stasis), leads, as in bronchitis, to the formation of a multitude of effete cells, and a consequent blocking-up or stuffing of the tubes, with abortive and disintegrated epithelium and fibrinous plugs. This is the large smooth kidney of acute albuminuria.

The second represents the granular degeneration of the kidneys. The granular kidney, either the larger or the smaller contracted type, results from disease of the fibrous tissue. The kidney, at first large from increase of the fibroid substance, becomes eventually smaller by its contraction if life lasts, and presents, Dr. George Harley, Dr. Dickinson, and others think, a form of disease closely allied to cirrhosis of the liver. The morbid change has its seat in the intertubular fibrous tissue, and is confined to the kidneys. While the third form included under the class of Bright's disease, is the amyloid, and is of a more widely diffused character. It has its origin in the minute blood-vessels; and the kidney, like other glandular organs, becomes infiltrated with a peculiar glistening material which is poured out from the small arteries, and eventually becomes diffused through the whole tissue. This glistening spermaceti-looking substance exhibits remarkable and characteristic reactions with iodine. The tissues wherein it is present being tinted of an orange-brown colour by an aqueous solution of that substance.

This arrangement has the decided merit of simplicity, and is, moreover, founded on observation of the facts and conditions which the microscope reveals. Examples typical of these very distinctive morbid processes are, without doubt, not infrequent in our post-mortem examinations; but the difficulty with me is to reduce all or every example we meet to one or other of these forms.

For instance, how constantly is what Dr. Dickinson calls tubal nephritis mixed up with, or followed by, granular degeneration, which results from an essentially different morbid process, according to this observer. Nevertheless, pathological science, as well as the students of renal diseases, must ever acknowledge the value of the researches of Dr. Dickinson. The multiplication of names for disturbed or diseased action based on the pathological views of an individual observer is to be regretted. Thus, for acute morbus Brightii we have already the following synonyms:—Albuminous nephritis, acute inflammatory dropsy, acute desquamative nephritis, and, lastly, tubal nephritis. This nomenclature is of little consequence to those devoted to renal pathology, for they are well acquainted with the individual views and opinions of the authors who have proposed their use. But to the student and even young practitioner this variety of designation for one and the same disease is, not to say puzzling—at least, often disheartening.

It signifies little, perhaps, which term is employed provided the application of it is uniformly to a definite group of symptoms accompanied by equally well-defined structural changes.

It is thus seen that the pathological conditions of the varieties of kidney found after death in albuminuria are closely allied. They may be conveniently classed as types of acute and chronic disease, or rather of inflammatory and non-inflammatory action, or of acute and chronic degeneration. There may be evidence of the coexistence of both acute and chronic products in the same organ. Thus, in the inflammatory type there will be an augmentation of volume either through the inflammatory congestion, or by the infiltration of the organ by an inflammatory albuminoid product, chiefly interstitial, with great increase in the formation of abortive and granular epithelium; while in the chronic form there may be also increase of volume, but now arising from the slow development of the fibroid element arising chiefly from long-continued passive congestion with great alteration in the character and appearance of the renal epithelium, which is highly fatty and granular.

These are the two representatives of acute and chronic albuminuria. But the large granular kidney of the chronic form often exhibits traces of concurrent inflammatory action. For there is frequently observed in these granular kidneys the presence of the same albuminoid material, only in less degree, which is the type of the acute disease. Such cases will in their clinical history often reveal the fact that during the progress of the chronic disease renal engorgement has occurred with traces of hæmaturia trifling in degree, accompanied by some transient feverishness, the sediment presenting fibrinous blood-casts, or isolated blood-corpuscles only, in addition to those which have hitherto marked the true chronic condition. These cases, I believe, never terminate in the granular red contracted kidney. It is supposed, however, that this latter type of chronic albuminuria is the sequel to the large granular kidney. Certainly both forms have many textural conditions in common.

They are both granular on the surface, both exhibit a large increase in the fibroid element, the contracted granular kidney the most so. In both there is hypertrophy of the walls of the small renal arteries, with corresponding hypertrophy of the heart. In both the epithelial cells are fatty and degenerated; but in the contracted kidney they have been swept away and have disappeared in great part, the tubes being empty and denuded; while in the large granular kidney they fill the tubes, though fatty and degenerate.

In the clinical history of *some* cases of the contracted granular kidney there may be recorded the pre-existence of dropsy with symptoms more or less characteristic of the acute form of disease. Such a fact gives favour to the opinion that the contracted kidney is but the sequel to the large granular kidney, or the large smooth white kidney. If every case of contracted kidney presented such a record of having passed through a stage of acute inflammatory action there would be an end to the controversy, and it would be unhesitatingly admitted that the contracted kidney was but the sequel to, or the latter stage of, the large granular or, the pale smooth kidney. But in my experience this is not so.

The cases of the small red granular kidney furnishing evidence of pre-existing acute disease of these organs are not numerous. The acute forms are more frequent in early life, the chronic in the middle period. The early stages of the acute form are more amenable to treatment than the chronic. The latter, which represent the true granular and fatty degeneration of the kidney, are incurable, or at best susceptible of amelioration or retardation.

§ 1. CHRONIC ALBUMINURIA *after Scarlet Fever.*—The albuminuria after scarlet fever has been treated as an acute disease, and in the majority of cases it is so. But Dr. Herman Weber has communicated to the Royal Medical and Chirurgical Society* some cases of albuminuria which have followed scarlet fever and other eruptive diseases at intervals sufficiently long after

* 'Med.-Chir. Trans.,' vol. xlix.

the fever to suggest one of two conclusions. 1. Either the seeds of the secondary renal disorder are capable of being maintained undeveloped for a long period and then germinate, as it were, with but a sluggish activity into life; or, 2, the renal disturbance is determined by other causes than the precedent fever, which was an accidental coincident. In the first case the albuminuria did not declare itself till two months after the scarlet fever. The urine was daily examined, almost daily after the eruptive fever, and not a trace of albumen was detected. Eleven weeks after the scarlet fever, during which interval perfect convalescence appeared to have been established, the child having returned to school, his general health began to fail. There was no fever, however, nor lumbar pain, nor any dropsical swelling; but the urine was now loaded with albumen, and the sediment contained granular tube-casts. Under appropriate treatment the urine subsequently lost all trace of albumen. A few months afterwards, however, from probable exposure to wet, albumen was for a short time again present, but eventually again disappeared. The second case was of a fatal character, and terminated in waxy kidneys seven years after the scarlet fever. The albuminuria continued through this long interval, and the patient died from symptoms of uræmic poisoning. Dr. Herman Weber has some very judicious remarks on the relation of those cases to the antecedent eruptive fevers. He says " The question offers itself whether, amongst the greater number of cases of Bright's disease, the origin of which is inscrutable to us, a certain proportion may not originate during the convalescence from eruptive fevers, and that thus these diseases belong to the remote causes of chronic renal affections ?"

Erysipelas and enteric fever may be regarded, also, as frequently precedent conditions to chronic renal disease. And the history of many cases of chronic disease of the kidneys which have come under my notice contains the record of an antecedent attack of enteric fever, just so far removed from the development of the renal symptoms as to justify the inference that the deteriorating effect of the fever on the blood was essentially the predisposing cause leading to renal degeneration in the form of chronic morbus Brightii. It will be seen, on reference to the alleged causes of non-inflammatory albuminuria, that all are characterised by a marked depreciation of the healthy qualities of the blood. Diphtheria is a frequent cause of chronic albuminuria.

§ 2. THE STRUMOUS TAINT.—Scrofulous degeneration of the kidney has long been recognised as a distinct pathological fact. This must not be confounded with tubercle of the kidney, although tubercle is recognised as the result of a strumous diathesis. There is a peculiar form of fatty degeneration of the epithelial secretory cells of the kidneys first distinctly noticed as the product or result of the scrofulous diathesis by Mr. Simon,* who artificially produced in animals a fatty degeneration of the kidneys by exposing them to conditions calculated to develop the strumous cachexia. Very numerous are the patients which our hospitals supply of examples of fatty degeneration of the kidney which is found in connection with some form or other of scrofula. Cases of phthisis frequently exhibit the most advanced degrees of granular and fatty degeneration. In some cases the amyloid form of degeneration prevails.

It may be said generally that whatever tends to deteriorate the vital energies acts as a predisposing cause to these chronic forms of renal degeneration. Thus may be explained the frequency of the disease among the inhabitants of overcrowded, ill-drained districts, who are ill-fed, intemperate, and dissolute; among whom the worst forms of disease are generated; where epidemics are most fatal; where enteric fever is rarely absent; and where cholera and other epidemics secure an abundant list of victims.

The susceptibility of individuals of a strumous diathesis to suffer renal disturbance on exposure to cold and wet, or to any of the exciting causes of renal disease, has been already noticed at pp. 22, 23, under the head of acute morbus Brightii. Not

* 'Med.-Chir. Trans.,' vol. xxix, p. 15; vol. xxx, p. 160.

less obnoxious is this form of constitution to the effects of chronic albuminuria, accompanied by pathological changes in the renal structures of a special and characteristic type. But this chronic form of disease is insidious in its origin, slow in its progress, and accompanied, as it usually is in such constitutions, by disease of other organs. It may be, and is frequently, overlooked unless some prominent symptom, such as dropsy, or anasarca of the extremities, or the clinical care of the practitioner, has directed attention to the urine, when its albuminous condition and the character of the casts seen in the sediment sufficiently announce the renal complication.

Phthisis is occasionally associated with that form of structural degeneration of the kidneys which is commonly called amyloid. But the young practitioner must carefully recollect that this change of structure in the kidneys is not tubercular. When tubercle is freely deposited in the lungs and is slowly passing through its fatal changes, it is quite exceptional to find tubercle in the kidneys.

On the other hand, chronic phthisis is most frequently associated with a peculiar form of fatty degeneration of the kidneys, sometimes purely fatty, at others clearly amyloid, and the connection of both forms of degeneration with the scrofulous constitution is now generally recognised.

Cancer, or the cancerous diathesis, also materially favours and is a precursor of the same form of chronic renal degeneration, the amyloid. It is now believed that amyloid disease is solely derived from some antecedent pyogenetic disease.

§ 3. THE SYPHILITIC TAINT.—The connection between constitutional syphilis and chronic or non-inflammatory albuminuria is somewhat more difficult to trace. It is certain that the impregnation of the constitution with the syphilitic virus, developing itself in the secondary or tertiary forms, is accompanied by marked deviations in the general nutrition of the body. A peculiar cachexia is as marked as in scrofula or cancer. The best agents of nutrition, the most carefully prescribed diet, fail in reparative power; and the sufferer from constitutional syphilis, more particularly if a dissolute and intemperate life, as is most frequently the case, be added to the list of deteriorating agencies, becomes obnoxious to the slow and insidious ravages of chronic albuminuria.

But it is chiefly in the offspring of syphilitic parents, in the miserable objects of hereditary or congenital syphilis, that the most marked examples of chronic renal degeneration are found, although the kidneys are perhaps less frequently affected than the liver.

§ 4. THE GOUTY TAINT.—Under the section on gouty nephritis (p. 27), the student has been cautioned, in a footnote, carefully to distinguish between those diseases of the kidney which may be excited by uric acid, gravel, or calculus, so common in the gouty constitution, and that form of shrivelled or atrophic kidney which has by common consent, since Dr. Todd's time, been designated as the atrophic or gouty kidney. As it does not follow that every gouty person must suffer from lithiasis, gravel, or calculus, although many do, so also it does not follow that a gouty patient should necessarily, in the sequel, become the subject of atrophy or wasting of the renal organs. Nevertheless, one or other of these morbid conditions is so often observed in connection with the gouty constitution, that the term gouty nephritis and the gouty kidney are intended to express different forms of diseased action apparently having a common origin. Gouty nephritis has already been described as occurring in a kidney suffering congestion or irritation from lithic acid, sand or gravel in a gouty subject. The atrophic, gouty or contracted kidney is pathologically very different. I recognise two forms of contracted kidney—the granular, red or yellow, contracted kidney, and the nodular, tuberculated, shrunken, or atrophic organ. The first may occur associated with gout, but chiefly with gout and some other deteriorating cause, particularly lead poisoning. It may also be traced to other causes, intemperance, an alcoholized blood, and whatever may dispose to chronic albuminuria. It is by some observers considered the sequel to, or the latter stage of the large granular kidney. The nodular contracted kidney I trace entirely to gout and its sequence, in deposits of urate of soda in the fibrous structures. Among gouty subjects, those in whom the gouty element has reached that point in which urate of soda has been freely deposited in the fibro-cartilaginous textures, are those in whom I believe the atrophic or contracted red nodular kidney (the gouty kidney) exclusively occurs.

§ 5. Reference has already been made (p. 13) to the action of the salts of lead and phosphorus on the renal organs, and it is there stated that these poisons do not appear to exercise any direct agency on the kidney. They do not produce nephritis nor any symptoms of acute action. But the impregnation of the system with either leads eventually to a chronic degeneration of structure, exemplified most frequently in the granular red kidney.

The association of lead impregnation with gout is also there alluded to; and it is in cases of paralysis from lead, in whom gouty symptoms develop themselves, that this chronic form of renal degeneration most usually occurs. It is not easy to trace the connection of these diseases, gout and lead impregnation, the one with the other. Which is the precursor? In the few cases from which I have been able to obtain a reliable history, the lead poison has been the forerunner. In each of these, however, the habits of the individual were just as likely to induce gout as the lead locked up in his system. In all, they were beer or spirit drinkers; in many, both. In every case there was the usual aspect of a broken-down constitution; a very marked earthy aspect of the skin; the saturnine margin along the alveoli predominated in all, and the general effects of the lead poison were more noticeable than those of gout. As usual, their connection with the painter's trade commenced in early life—with their apprenticeship, so that for many years the deleterious effects had been accumulating in the system. Lead is stored up in the organism in combination with albumen, from which it can be set free only by the agency of some few remedies; by the iodide of potassium especially. The kidneys are capable of excreting the salts of lead, and it is well known that this metal can be detected in the urine of those suffering from the accumulation of lead in the system, even before any remedial agent has come into operation. May not, therefore, the passage of such an agent through the renal organs lay the foundation for those changes of structure which subsequently develop into a chronic granular degeneration?

§ 6. INTEMPERANCE *through an alcoholized blood.*—Intemperance brings more victims to chronic renal diseases than all the other causes put together. Not only is it the probable parent of many fatal cases of acute disease of the kidneys, but it is palpably and notoriously the cause and source of the majority of cases of chronic albuminuria in adult age. Something may be due to the over-stimulation of the renal organs by the direct action of alcoholic stimuli; but the degeneration of the kidneys mainly, if not solely, originates in that deterioration of the nutritive qualities of the blood which is so palpably defective in the victim of intemperance. The aspect of these individuals is most marked. If the cheeks are red it is not the ruddy colour of health, it is a dusky tint of blood defectively oxidised. In some there is a certain yellow pallor with a peculiar mottling or marbling of the cheeks by fine capillary injection. To this may be added a dull and leaden eye, from which all vivacity is banished; such is the aspect of the habitual toper. These individuals are not in their own sense drunkards; they never drink to lose their senses, although their intellect may be, and is, in a perpetual muddle.

Thus, the habitual toper is never free from the spurious excitement of alcohol; his nervous energy is kept to a certain degree of tension, beyond which it rarely passes, but which has destroyed the elasticity of the nerve power; for let the stimulus be withdrawn or discontinued, the nervous system relaxes like an overstretched cord and becomes enervated and listless.

The loss of equilibrium in the nerve force is not a solitary effect of the alcoholized state of the system which the abuse of stimulants brings about. Important functions fail in action. Morning retchings, gastric catarrh, hæmatemesis, precede a total loss of appetite, and mark the initiation of disease. Thus the stomach and liver are oftentimes the first important organs which declare the poisoning influence of alcohol and intemperance. With a failing appetite comes distaste for solid food, and speedy rejection of what little may be taken; nutrition becomes defective, the blood deteriorates. Alcoholic stimulants, taken in such amount as to be in excess of what the lungs can eliminate as carbonic acid and water, load the system with the results of imperfect oxidation. An impure blood, thus loaded with carbonaceous products, becomes the ready parent of those transformations of tissue which have fat and cholesterin on the one side, or albuminoid or fibroid indurations on the other.

The liver becomes fatty, the blood-vessels become atheromatous, the small arteries of certain organs, particularly the kidneys, become thickened, and their channels narrowed, fibrous and albuminoid elements accumulate, the gland-cells become fatty and effete, and there is eventually developed in the kidneys that form of granular and fatty degeneration which is the offspring of intemperance.

The large granular kidney is the most frequent result of these habits, but the small contracted kidney is occasionally met with as the sequel to the same cause.

§ 7. CHRONIC ALBUMINURIA, *the sequel to obstructive conditions in the circulation.—Valvular diseases of the Heart.—Emphysema, and Chronic Bronchitis, and occasionally disease of the Liver.*—In the early period of these disorders the kidneys rarely suffer. It is only when the obstruction in the circulation has reached a certain limit, when dropsical effusion begins to show itself in the lower extremities, and the retardation in the venous current is felt in the inferior cava, that the circulation through the kidneys begins to be embarrassed in consequence of the return of blood from the kidneys through the emulgent vessels not being freely delivered through the vena cava. The first effect of this embarrassment in the renal circulation is the presence of albumen in small quantity in the urine, unaccompanied, however, by any special microscopic object in the sediment. The albumen seems derived at first from a simple blood stasis. The venous plexus surrounding the uriniferous tubules suffers a condition similar to compression of the veins of a limb, with similar consequences. The vessels swell and the surrounding textures become infiltrated with the serous or watery element of the blood. This permeates the uriniferous tubes and adds an albuminous element to the otherwise natural urine.

That this is so seems proved by the total disappearance of the albumen from the urine in the early stage of these disorders, when treatment has successfully for a time combated the primary disorder and relieved the embarrassed circulation. It is not till late, and after a prolonged duration of these diseases, that the organic or structural changes in the kidneys, about to be described, begin to take effect, and the albumen to be accompanied by microscopic evidence of chronic disease of the kidney. The effect of mechanically induced congestion to produce induration and change of structure in such organs as the liver and kidneys * is a well-recognised pathological law. The impediment to the circulation causes an exudation of material from the blood, which, whether it be lymph, such as transudes in ordinary inflammatory stasis, or not, eventually leads to condensation and induration of structure, and microscopically is seen either as a fine granular matter or having the more organized character of fibrous or fibroid tissue. Thus, in disease of the heart and lungs, valvular disease of the former, and such disorders as emphysema and chronic bronchitis of the latter, the impediment to the circulation reaches or is felt in the right side of the heart, and the organs whose blood is more immediately or directly emptied into the cava are those which suffer most prominently. Among them the

* See 'Trans. Med. Chir. Soc.,' vol. xliii, a paper by Sir W. Jenner.

kidneys occupy a chief place. Partly because their nutrition suffers from the stasis in the intertubular plexus, and partly because, in those diseases in which the transmission of blood through the lungs is impeded, there must necessarily follow an accumulation in the blood of effete or sub-oxidized material in consequence of the imperfect respiratory function. The two forms of diseased kidney arising from these causes are the contracted yellow granular kidney, and the large granular kidney.

The young practitioner must not forget that these disorders of the circulation, whether arising in the lungs, heart, or liver, and the consequences as seen on remoter organs, especially the kidneys, are evidence of a wide-spread decay pervading the organism rather than of disease of any separate or single organ. The implication of the kidneys, as indicated by the albuminous urine, and granular and hyaline and other casts seen in the sediment, must be regarded as the last link probably in the chain of decay, for degeneration is everywhere evident. The chief textures of the body yield to microscopic research proof of departure from the physiological type of structure, and the conviction remains that the renal degeneration, although promoted by the obstructive conditions in the circulation, is nevertheless but a presage that the limit of life is approaching, and that these degenerations are but the expression of a certain special form of decay to which all organized beings are finally subject, modified as the form may be by temperament, constitutional proclivities, by habits of life, or concurrent disease.

§ 8. MORBID ANATOMY.—The structural changes in cases of albuminuria observed in the kidneys after death, whether by the unaided eye or by the microscope, differ very much one from another.

I cannot but think that a great deal of confusion and unnecessary perplexity has been introduced into the subject of the morbid anatomy of Bright's disease, particularly by the German school, by too great attention having been given to the physical characters of the kidneys, their size, weight, colour, hardness, softness, the aspect of the cortical surface, as regarded roughness or smoothness, and other characters estimated only by the unaided eye, to the neglect of those alterations which the microscope reveals in the tubular, cell and vascular elements of the diseased organ.* It is here that the pathology of Bright's disease owes so much to the labours of Dr. George Johnson. His observations and conclusions have been based on microscopic investigation, and, consequently, through him we know more of the morbid -anatomy of this disease than could even be learnt from the minute, but perplexing, descriptions of the continental schools with their eight or nine forms of the disease, based on certain physical differences estimated or viewed chiefly by the unaided sense of sight.

To the name of Dr. George Johnson must now be added those of Dr. Grainger Stewart and Dr. Dickinson, whose microscopic details of the morbid anatomy of Bright's diseases, particularly of the amyloid form of degeneration, have cleared much that was confused, and have finally established the pathology of these varieties of renal disease.

The three types of [alteration, by an inflammatory process, are here repeated for the sake of comparison with the structural changes brought about by the non-inflammatory or more chronic forms of the disease.

1. The kidneys exhibit a state of intense congestion or active hyperæmia; purplish red, or brownish red; swollen; the Malpighian bodies and renal tubes charged with blood and fibrinous blood-casts from rupture of the Malpighian vessels; the pyramids are deeply striated with a purplish colour; by pressure a fluid is squeezed from the straight tubes, which consist of fibrinous blood-casts, coarse granular casts, and free blood-corpuscles.

2. In the second variety the kidneys are increased in weight, of a yellowish or yellowish flesh hue. The cones deeply striated. The tunic when torn off shows the cortical surface

* On this point I regret to differ from Dr. Roberts. See 'Urinary and Renal Diseases,' p. 297.

mottled with red patches. The cortex is thickened, and the distance between the base of the cones and the surface increased. The tubes are filled with fibrinous and granular matter, in which are seen numerous detached epithelial cells. The tubes also contain large numbers of the compound granule cell (Gluges' corpuscles). The Malpighian bodies are distended and filled with a granular or fibrinous material. The fluid obtained by squeezing the cones, under the microscope consists of casts containing epithelium in great abundance (granular epithelial casts), here and there probably a blood-corpuscle imbedded. A good many free epithelial cells, cloudy, opaque, and granular; sections under the microscope show fibrinous coagula in the tubes of the surface. This form represents probably the earlier stage of the pale smooth kidney.

3. In the third form the kidneys are much increased in weight and size, weighing often from eight to even sixteen ounces each. The colour is palish yellow, or pale flesh tint, or even white. Its appearance is that of a bloodless organ. The tunic is easily removed, when the surface is seen smooth, with here and there some stellar-like spots of vascularity. The cortical layer is much increased in thickness by a fine granular material which invades every part of the structure, Malpighian bodies, convoluted tubes, and intertubular spaces. The tubes appear choked, it may be said, with epithelial cells, surrounded or imbedded in an amorphous, pale granular material, with more or less fatty débris scattered through it. The walls of the small arteries throughout both kidneys are thickened and hypertrophied. Pressure on the cones produces a fluid which contains an abundance of renal epithelium, of faintly granular casts, containing abortive epithelial cells with many resplendent granules—the nuclei of broken-up cell structure. The large compound granule cells are also numerous, and are either free or contained in a cast. The contents of these large cells vary from a fine granular appearance to clusters of highly resplendent nuclei, with or without the cell-wall. In the latter case they may fancifully be likened to grape clusters or the mulberry fruit, and hence

sometimes called *botryoidal*. This microscopic examination of the fluid squeezed from the apex of the cones becomes a valuable aid to diagnosis. The fluid squeezed from the straight tubes contains the same objects as those which appear in the sediment during life; it is washed out by the urinary current, and forms the sediment when the urine is set aside to rest. If this sediment, therefore, during life exhibits similar microscopic characters to what is squeezed from the cones in other cases, we may rationally infer that the kidney which yields these objects is in similar condition. So that when the sediment exhibits fibrinous blood-casts, coarse granular casts containing blood-corpuscles, the stage of engorgement or early inflammatory action is known to be present. If the sediment consist of epithelial casts more or less granular, with large granule cells, we infer that the second form is present. If we find in the sediment casts of both large and small diameter, slightly or sparsely granular, with many scattered resplendent nuclei, abortive epithelial cells, with many large granule cells, with clusters of highly refractive nuclei, we correctly infer that the kidney has advanced to the stage of the large anæmic variety, and that the termination of the disease must be fatal.

Between this third variety, as the result of an acute process, and a form of disease associated with chronic Bright's disease, a certain degree of similarity exists. In some cases of a protracted character, having moreover no inflammatory origin, the kidneys are found large, anæmic, mottled, and *granular*. The surface when the capsule is removed is covered with granulations, very different from the smooth surface of the pale anæmic kidney found after scarlet fever or acute morbus Brightii; yet these latter cases often run a course of months' duration. In that sense the disease is chronic; but the increase of size and weight is clearly traced to inflammatory action, the infiltration of the whole organ with an albuminoid or granular and fatty exudation. This large, white, smooth kidney is the type of the acute form of Bright's disease, while the large, pale, mottled, granular kidney is one of the varieties of the chronic form, and is the result of a different pathological process, or rather is the effect of a combination of two morbid processes; the one a passive, insidious, but permanent state of congestion, such as may be produced by obstructive disease in the heart or lungs, leading to increase in the fibrous interstitial element of the organ; the other, accessions of active or inflammatory congestion, developed by the common exciting causes, cold or wet, and followed by the ordinary exudative product, the fine, albuminoid granular material which makes up the bulk of the smooth anæmic kidney of the acute disease. This large, mottled, granular kidney, with its accompanying diffuse dropsy, represents, therefore, not a single pathological process, but a double or complex one. It is a well-recognised clinical fact, that some cases of chronic albuminuria display during their progress occasional attacks, either of hæmaturia or of urine containing proof of a congestive condition of the kidneys, in the shape of isolated blood-corpuscles, granular casts containing blood-corpuscles indicative of blood passing through the renal tubes. Such cases of chronic albuminuria at the outset cannot be traced to an inflammatory origin; they usually commence most insidiously with varying symptoms of failing or decaying health, sometimes preceded by cardiac or pulmonary disorder. The patient is often able to continue his occupation in a moderate way; the urine is, however, albuminous. There is but little dropsy at first, not sufficient to incapacitate from business, provided it be not heavy or laborious. Usually after some accidental exposure or neglect of care, the patient suffers an accession of acute symptoms, such as slight feverishness, or perhaps only the symptoms of an ordinary catarrh, accompanied, however, by evidence, already mentioned, of an accession of low inflammatory action in the kidneys. This may happen more than once in the course of chronic disease; and it appears to me that these attacks, grafted on a kidney already the seat of disturbed action, produce in a low degree the results found in the kidneys to follow the more energetic, acute process; and the ultimate result is a kidney increased in size and weight, anæmic, and offering, through the microscope, evidence of an infiltration of exudative material, both inter- as well as intra-tubular, granular and fatty, and in all respects, save in amount and extent, similar to what is found in acute albuminuria.

I therefore think the granular, pale kidney is the result of several minor attacks of chronic inflammatory action, occurring in an organ already the seat of the morbid changes incidental to chronic albuminuria. It is, therefore, essentially a chronic form of the disease with intermittent inflammatory action.

It is on this account, namely, the occasional supervention of a state of congestion of the kidneys in the progress of the strictly chronic form of the disease, which renders it so difficult to reduce many examples of renal disease (Bright's disease) to this arbitrary classification of acute and chronic. Nevertheless, there are certain advantages in adhering to this arrangement, if we keep in view that the one class commences in inflammatory action, and yields evidence on examination of an excess of granular albuminoid deposit, the produce of such disturbance; while the other class has no such origin, although in its course congestive conditions may now and then be excited, but from which the disease does not date its commencement, and are but accidental interferences with the original or fundamental disturbance.

The structural changes most characteristic of the chronic forms of albuminuria are typified in the following four varieties:

I. The granular contracted red kidney, colour reddish-yellow, or yellowish with characteristic granulations.

II. The granular enlarged pale mottled kidney, with similar granulations.

III. The amyloid, waxy, or lardaceous kidney.

IV. The atrophic, gouty, or contracted kidney.

I. THE GRANULAR CONTRACTED KIDNEY.—Its colour varies very much; it is often reddish yellow. The chief characteristic, however, of this variety is seen in the contracted or condensed state of the cortical layer, covered externally with yellow sand-like or coarser granulations. This form is considered by many observers to represent a stage of contraction subsequent to one of enlargement. It is this variety which has supplied the ground for the belief that a shrivelled and contracted kidney may succeed to, and be the terminating stage of, the large anæmic kidney. This form of diseased structure has, moreover, by some writers been confounded with the atrophic contracted kidney which I have placed fourth in the series of examples of chronic albuminuria; both the causes and the nature of the structure, I believe, materially differ. Dr. Geo. Johnson in a paper published in the forty-second volume of the 'Medico-Chirurgical Transactions,' expressly distinguishes between the two, and justly remarks that the appearance presented by a contracted fat kidney, with its characteristic small yellow granulations, is very different from the small kidney which has been the subject of chronic desquamative " disease;" this form is red and nodulated, or even tuberculated, on the surface, and the organ more distinctly shrivelled or contracted than the so-called contracted fat kidney. Dr. Geo. Johnson, in the paper just quoted, considers the contracted fat kidney, as he calls it, to have been over enlarged, and that this condensed condition of the cortical substance is caused by the destruction of the renal epithelium by disintegration, and the choking of the tubes by unorganized fibrine; conditions which are constantly associated with atrophy of the convoluted tubes.

The opinion that the granular contracted fat kidney is the sequel to the chronic enlarged kidney rests, in Dr. Geo. Johnson's view, on the clinical history of the case. He thinks that the early history of these cases of contracting kidney is that of chronic enlarged kidney; for he says, in some the disease originates in an attack of acute dropsy with albuminuria.

This granular contracted kidney corresponds to Dr. Bright's third form or stage.

It is but little augmented in volume and weight. The capsule is slightly adherent, and when removed displays a rough surface of fine yellow granulations. The chief characteristic of this variety is the thin and lessened diameter of the cortical layer, so that the base of the cones comes in close contact with the surface. The pyramids are of natural colour —red, or perhaps purplish red—and strongly striated. The fluid squeezed from the cones displays to the microscope numerous granule cells (Gluges), abortive and granular epithelial cells, with casts more or less granular; some are partially hyaline, speckled as it were with fat-granules, the débris of disintegrated cell structure. The tubes of the condensed cortical part, or what can be made out of them; some appear denuded and empty, others are filled with a granular and fatty débris. The fibrous element in the structure of the kidney appears increased, but whether this is due to the contraction of denuded and empty renal tubes, or from a fresh accession of fibrous tissue, the result of interstitial deposit, I am not prepared to say.

Dr. Dickinson, in a paper published in the 'Medico-Chirurgical Transactions,' vol. xliv, "On Diseases of the Kidney accompanied by Albuminuria," has described the morbid anatomy of granular degeneration with much perspicuity. He considers the granulations, whether coarse or fine, seen on the surface of all kidneys, the subject of granular degeneration, to arise from a formation of intertubular fibroid tissue. He says,* the new formation commences on the surface at a number of detached points, and presses inward among the tubes as separate processes; subsequently this fibroid matter contracts in a manner similar to what occurs in cirrhosis of the liver. Rokitansky on the other hand considers the granulations to be formed by an exudation of granular material into the Malpighian bodies.†

It must not be overlooked that, in the clinical history of these cases of renal degeneration, the accession of engorgement or active inflammatory congestion of the kidneys is not unfrequent. A patient suffering from chronic albuminuria, exposed to cold and wet, would manifest all the signs of this

* 'Pathological Transactions,' vol. xiv, p. 184.
† 'Pathological Anatomy,' vol. ii, p. 199. Sydenham Society Edition.

congestion, expressed by traces of blood in the urine, feverishness, and the manifestation of increased dropsy, with diminution in the amount of urine passed. It is the engrafting of those inflammatory attacks on a kidney previously disorganized by chronic disease that oftentimes creates the difficulty of separating the granular contracting kidney from the larger mottled granular kidney. Dr. Grainger Stewart has very clearly described, these blendings of the results of inflammatory action with pre-existing chronic disease.

In this form is seen a very characteristic change in the structure of the renal arteries, consisting of an hypertrophy and thickening of their walls, first noticed and described by Dr. George Johnson, and more recently illustrated by him in vol. li of the 'Medico-Chirurgical Transactions.' These kidneys, as well as the following, are examples of probably the most familiar form of morbus Brightii, that of the so-called granular and fatty degeneration of the kidneys.

II. THE LARGE GRANULAR AND FATTY KIDNEY; *often mottled*.—This kidney exceeds in weight the average. Its colour varies very much; it never possesses the extreme pallor or whiteness of the smooth kidney. It is sometimes of a yellowish flesh tint, and mottled with spots of a red injection. It is occasionally of a dirty grey, or yellowish grey, with a distinctly granular surface. The capsule is slightly adherent, and when removed exhibits the cortical surface granular in varying degrees—as fine as sand, or as coarse as salmon roe. These granulations appear to be caused partly by an increase of the fibrous tissues of the organ, and partly by infiltration of the Malpighian bodies, or by fibrinous deposit or exudation within them. The tubes also exhibit a process of disintegration or denudation, being filled with abortive cells, detached from the basement membrane and filling the tubes to repletion, conveying the idea of how much such a turgid state of the renal tubes must, by pressure on the intra-tubular vessels, embarrass and cause obstruction to the circulation, and help to lessen the amount of the urinary secretion and aggravate the dropsical condition of the patient. It is in this form that the urine remains so persistently scanty in contradistinction to the abundant secretion in the contracted variety, and diuretic remedies so constantly fail to increase its amount. This variety of kidney exhibits a greater proportion of fatty elements than any of the others.

III. THE AMYLOID, WAXY, OR LARDACEOUS KIDNEY.—This form of diseased kidney has very distinctive features. It may exceed, though only slightly, the average weight. In substance it is firm and hard to the knife. The capsule is easily detached, and the cortical surface usually has a smooth, glistening appearance of a yellowish tint, showing numerous points having a waxy or even spermaceti look. On a section being made the scalpel cuts as through bacon; hence the name used by the Germans to indicate this variety. This section displays the cones contrasting very much in colour with the cortex. The tubes of the cortex, as well as the straight tubes of the cones, are here and there marked by a glistening spermaceti-like deposit. Nor are the renal tubes the only seat of this change; the muscular coat of the arteries is infiltrated with it, their walls are thickened to a great extent, and their internal diameter proportionately lessened. The material that thus invades the tubular as well as the vascular structure of the organ differs altogether from any other product of diseased action found in the kidney. This material—this waxy, spermaceti-looking substance—exhibits some very peculiar and distinct reaction, with certain chemical agents or tests. As these reactions were thought to be similar to what occurs with vegetable cellulose it erroneously received the name of amyloid.

The fluid squeezed from the cones displays large and small oily casts, and epithelial cells in the highest degree of fatty degeneration. These oily casts are very characteristic. Some of the casts are also densely granular, and rarely yield to iodine the same reaction as the so-called amyloid substance found after death in the kidneys.

Dr. Dickinson has brought forward some very novel and ingenious views as to the nature and origin of the amyloid substance which forms one of the varieties of renal degeneration.* He considers this substance as de-alkalized fibrine. The morbid deposit first appears on the walls of the small arteries from which it penetrates and infiltrates the neighbouring tissues. Several observers have noticed the frequency with which this form of degeneration has occurred associated with exhausting diseases of a pus-draining character. Dr. Wilks, in an admirable paper in the 'Guy's Hospital Reports' of 1865, traces it to tubercular and syphilitic diseases. Dr. Grainger Stewart also, and more recently Dr. Dickinson, have established its connection with suppurative diseases. In the paper already quoted this latter observer has endeavoured to trace the relation which subsists between the deposit of new material from the blood and the drain of pus from the system in certain pyogenetic diseases. The chemical composition of pus is well known. The fluid part, the liquor puris, is an albuminous fluid deficient in fibrine, a large proportion of the alkaline salts of potash and soda giving it a marked alkaline reaction. Analysis proves that these alkaline salts exist in greater proportion in the liquor puris than in the serum of the blood. Hence it is inferred that a great discharge of pus is equivalent to depriving the blood and tissues of their due proportion of albumen and alkaline salts. Dr. Dickinson's next step in tracing the relation between amyloid deposit and a purulent drain is to establish the fibrinous character of the new deposit. "This new formation readily becomes converted into fibrous tissue. When seen in bulk it soon assumes, under the microscope, a fibro-nucleated structure, and its presence in small amount is evinced by a thickening of all the fibrous structures with which it comes in contact. After its deposition it undergoes a regular process of contraction, as is evinced by the change which takes place in the parts involved."

Now, this new material, essentially fibrinous in its origin,

* "On the Nature of the Waxy, Lardaceous, or Amyloid Deposit." 'Med.-Chir. Trans.,' vol. l, p. 39.

exhibits towards iodine, or its aqueous solution, a remarkable and very characteristic reaction. It becomes stained of an orange or reddish-brown colour, healthy texture receiving only a yellow colour, and ordinary fibrine exhibiting no special reaction with iodine differing from healthy tissue. This establishes the fact that iodine displays a special agency towards this new material only. And this is further proved by the fact that this property is not destroyed by boiling or macerating in either water or alcohol, or by macerating in strong or weak solutions of acids. Moreover, Dr. Dickinson has shown that the property is retained by this substance for years, although kept in alcohol or methylated spirit; so that amyloid may be determined to exist in preparations that have been lodged for years on the shelves of the museum, so tenacious is this substance of its iodine affinity. But, nevertheless, there are agents which will instantly destroy this characteristic affinity. Digest the amyloid substance either after the iodine test has given the orange-brown colour, or before the test is applied in either of the fixed alkalies or their alkaline salts, and the colour instantly disappears if the test has brought out the special tint; or it fails to act at all if the amyloid substance has first been moistened with an alkali or an alkaline salt. Moreover, if this substance be once alkalinized it can never be made to yield its peculiar colour with iodine; so that neither washing nor digestion in acids can restore this singular reaction. It would thus seem proved that the iodine test depended on the absence of one or other of the fixed alkalies, potash or soda, or their alkaline salts.

There is yet one other proof resting on the chemical properties of alkalies to decolorise the salts of indigo. If sulphate of indigo be applied to almost any animal texture the original colour of a deep indigo blue slowly fades and passes into a palish green. This change of colour is produced by the alkalies which are contained in healthy tissues. But when the solution of indigo is applied to any organ containing amyloid, those portions containing the morbid deposit retain the deep blue colour, the other parts free from this deposit being slowly decolorised from the action of the alkaline salts present in them. Dr. Dickinson further adduces the analysis of certain healthy organs and tissues, as well as some in which the amyloid deposit existed, which certainly establish the fact of the deficiency of alkali, either potash or soda, in amyloid substance when compared with healthy structure.

The argument may be thus summed up—

1. *Anatomically*.—Amyloid, is modified fibrine; it has a fibrinous origin, and it has all the properties of fibroid tissue.

2. *Chemically*.—Amyloid, when acted on by solution of iodine or sulphate of indigo, displays and retains properties of colour which the surrounding textures do not exhibit. This colour is instantly discharged or destroyed by the addition of an alkali.

On healthy fibrine neither iodine nor sulphate of indigo produce any effect different from other healthy tissues. But fibrine digested in dilute hydrochloric acid becomes deprived of its alkali, and then behaves to these re-agents in a manner similar to amyloid.

Chemically, therefore, amyloid is de-alkalized fibrine.

Now, amyloid is found chiefly in those who have suffered from some exhausting disease, in which a long continued purulent drain has deprived the blood of its ordinary proportion of alkali and alkaline salts. Pathologically we know that blood deficient in these exhibits a peculiar aptitude to part with its fibrine in the form of morbid deposit; and thus is established a certain logical connection between the subtraction of alkali from the blood, in pyogenitic disease, and the deposit of a material deficient in that property, and possessing as above shown the characters of a dealkalized substance. Dr. Dickinson proposes to apply the term depurative infiltration to this deposit, and depurative disease as signifying this class of degeneration hitherto called amyloid, waxy, and lardaceous. Dr. Dickinson has earned the respect of all pathologists as well as practical physicians by his having so completely dissipated the erroneous analogy supposed to exist between this deposit and vegetable starch or cellulose. He has given this substance its true position among morbid products; and if we may hesitate to adopt the term depurative suggested by him, it is from no inferior estimation of the success with which he has pursued this inquiry, or the conclusion at which he has arrived as to the chemical and anatomical character of this deposit, but rather from an opinion that a better term may be selected which shall simply express some leading property of the substance, and without implying any theory as to its origin or cause of formation.

IV. THE ATROPHIC, GOUTY, OR CONTRACTED NODULAR KIDNEY.—The fourth form is that of the shrunken, shrivelled, contracted kidney. It is less in volume and weight than the average. The capsule is strongly adherent, and with difficulty torn from the cortical part, which then exhibits a tuberculated or nodulated surface. The colour varies from a red to a reddish brown. The substance of the organ cuts firm. This section shows the cortical substance reduced to the thinnest lamina. The base of the cones or pyramids seem almost in contact with the surface. The cones have a red, not unhealthy appearance; they are often striated with white opaque streaks, which on further examination prove to be urate of soda. The blood-vessels reach the highest degree of thickening; and many are completely obliterated. A microscopic examination shows the tubes of the cortex to be in great part empty, flattened, and condensed. The Malpighian bodies are filled with a few abortive and fatty epithelial cells.

A section of the denser part of the cortex at the base of the cones displays a fibrous-looking structure, having something of the appearance of tubes which have become atrophied and condensed. The tubes which have still preserved their form, appear empty and deprived of epithelium. The fibrous tissue is indurated or appears so, and may be the result of the contraction and shrivelling of the denuded renal tubes.

This variety is occasionally confounded with the granular

contracted kidney, and termed the granular, red, or cirrhotic kidney, which latter form is, as has been before mentioned, supposed by some to be the sequel to the enlarged granular kidney. Between the two there is the greatest possible distinction, both as regards symptoms during life, and the structural condition of the organ revealed by the microscope.

Wedl,[*] in his 'Pathological Histology on the Subject of Atrophies,' and speaking of the contracted nodular kidney, says this form of renal atrophy should be carefully distinguished from that accompanying Bright's disease, and which originates either in an exudative process, or frequently in a new formation of connective tissue in the interstitial substance of the renal parenchyma. Of this contracted nodular kidney, he says, the surface is nodulated, the consistence of the organ dense. The rounded elevations on the surface are formed by the shrunken tubuli uriniferi collected into groups, forming sacculi or pouches filled with a fine granular matter, fatty oftentimes. Renal sand or minute calculi are seen in the vessels, and diffuse atheromatous deposits are discovered in the larger arteries and even in the capillaries themselves. This form of atrophied kidney has been observed in extreme old age, and though apparently unconnected with gout, in some cases arises from the deposit of gouty material, the urates of soda and ammonia, as well as uric acid in the tubules, as well as in the interstitial fibrous structure of the kidney.

The following appear to me to be the chief reasons for distinguishing between the granular red contracted kidney and the atrophic nodular gouty kidney.

First, as to symptoms. Dropsy is not a prominent or significant symptom in either, yet it may, and does, occur sometimes, if only to the extent of anasarca of the lower extremities in the granular cirrhotic kidney, while it is rarely present in the atrophic gouty form. In both forms the urine is abundant, pale, of low specific gravity, albuminous, and deficient in urea.

The granular red cirrhosed kidney occurs either as the sequel

[*] 'Pathological Histology,' Sydenham Society Edition, p. 169.

to, or accompaniment of, other diseases, such as those produced by intemperance, by gout, by lead poison; or the history may show no such antecedent, and the disease may arise without any marked cause, while the atrophic nodular kidney has no other source or origin than gout, or rather the results of the gouty deposit in the textures of the organ. For this atrophy of the organ arises only from the deposit of urate of soda in the tubes and fibrous matrix of the organ, with a subsequent destruction and atrophy of the cells, with shrinking of the empty renal tubes, and eventual contraction of the whole organ.

The two forms of contracted kidney differ very much in colour. The granular contracted kidney is yellowish red, dusky, greyish yellow, with a fine granular surface like sand, or coarser. The atrophic, or gouty, contracted kidney is red, reddish-brown, dense and firm in substance; the capsule with difficulty removed, and the surface nodulated or tuberculate. These external characters sufficiently distinguish the one from the other. With respect to the process by which this atrophy or contraction of the organ is brought about, it appears to me to differ the one from the other.

The granular red contracted kidney may owe the condensation and induration of its substance to an increase in the fibrous element, the result of interstitial deposit arising from mechanical impediment to the return of blood through the emulgent vessels, and a subsequent shrinking or contraction of it, and if so, the process becomes analogous to what takes place in the liver in cirrhosis. This contracted kidney has been by some writers called the cirrhosed kidney;* and Dr. Harley describes the contracted granular kidney as the result of frequent attacks of inflammation, and that all three forms of enlarged, fatty, and waxy kidneys are liable to shrink and become cirrhosed by a process similar to what takes place in the liver. These are the views of the German school of pathology.† I cannot concur with these views, so far as relates

* 'On Albuminuria,' by Dr. George Harley, p. 29.
† See a very good summary of the views of the German, French, and

to the doctrine that the granular contracted kidney is the sequel to, and latter stage of, the pale smooth kidney of the acute form, or the enlarged granular kidney of a more chronic form of disease.

It is certainly an interesting pathological object to determine how far these seven varieties of diseased kidney are related to each other, or whether they represent progressive steps or stages of one morbid process, or whether they are essentially different and distinct. The first three forms, representing the acute process, may be considered as examples of different stages or different degrees of intensity of one morbid process, commencing in inflammatory engorgement, followed by an interstitial exudation of a granular or albuminoid material, as well as a large development of abortive, renal epithelium, which chokes the tubes and adds to the embarrassment of the circulation through the organ.

The large, smooth kidney is the result of this process being prolonged beyond the stages of engorgement and stasis. Months may pass without any, or but trifling, mitigation of symptoms. The duration of the disease may justify some in viewing this as an example of chronic rather than acute disease. But, as I have previously announced, the disease proceeds uninterruptedly from the inflammatory stage to this last final one; and this result of diseased action I will therefore place among the acute forms. These three forms, typical of acute disease, may be viewed as examples of the same morbid process at different stages or acting with different degrees of intensity. This acute disease is in its early stage more manageable than the chronic; and the cases which reach the fatal stage of the large, smooth kidney are small in comparison with the number of those who suffer the primary symptoms with its accompanying dropsy, and eventually recover.

The first two forms of the chronic class of albuminuria, so far as regards the fatty degeneration of the renal epithelial cells, have that in common with the acute form of the disease.

English Schools of Pathology on "Morbus Brightii," by Dr. J. Abeille, 'Traité des Maladies à Urines Albuminenses,' p. 93, et seq. Paris, 1863.

Dr. Dickinson, indeed, does not consider this fatty degeneration of the cells as indicative of a morbid process. I differ from him, however, on this point. This fatty degeneration of the renal gland-cell is of the very essence of the disease, and in these chronic forms seem to me to constitute a very distinctive feature. In these chronic forms, except in the large granular kidney, the tubes are not choked with abortive, cloudy, granular, epithelial cells, filling and apparently distending the tube as in the acute disease. The cells are notoriously loaded with fat-granules. There is an abundance of these fatty, degraded cells, but they do not overload the tubes. The tubes now and then appear altogether denuded of epithelium, and empty. This is particularly the case in the contracted granular kidney. The chief feature is the excess of fibroid structure, from the subsequent shrinking or contraction of which the contracted granular kidney is supposed to be formed. The distinction, then, between the two forms of albuminuria, the acute and chronic, appear to me to rest on the different character of the interstitial deposit.

The engorgement of the kidneys by inflammatory action leads to its usual results, as witnessed in other organs, an infiltration of the structures, both interstitial and tubular, with the material derived from the blood, fibrinous and albuminoid. In the tubes it appears as fibrinous coagula, in the interstitial structure it assumes the appearance of an almost colourless granular material. These exudations, whether as coagula or casts in the tubes, or as an interstitial infiltration, usually present this granular aspect. Dr. Dickinson thinks there is no interstitial deposit, and that the augmented volume of the gland arises from the immense proliferation of the epithelial cells and their distending to repletion the renal tubes. In chronic disease, the varieties which have no definite starting point, no inflammatory origin, although inflammatory agency may affect the organ during the progress of the disease, the product materially differs. There is, however, something of relationship between the first two forms in this chronic class, the large granular and the contracted granular kidney, and there is also some analogy between them and the pale, smooth kidney of acute disease. The deviation from healthy cell structure is in the same direction in both. The epithelial gland-cell degenerates alike in both. Fatty nuclei take the place of the isolated, well-defined nucleus, and there appears in both forms numerous compound granule cells, always evidence of deterioration in the nutrition of the textures. Thus far the analogy holds. But the structure of the blood-vessels, the small renal arteries, is materially changed in these chronic forms; less apparent, or altogether absent in the acute varieties. Moreover, the increase in the volume of the gland appears dependent on a greater development of the fibrous element in the large granular kidney; by the contraction of which the granular contracted kidney is supposed to result.

These structural changes seem to depend on a slow and insidious defect in the nutrition of the kidney, sometimes effected by such passive engorgement of the organ as obstructive conditions in the circulation through the heart or lungs may bring about; sometimes by a general deterioration of the nutritive qualities of the blood which may be caused by intemperance, by fevers, or by the local influences which generate them, and various other causes already enumerated. These five forms, then, of albuminuria, three representing the stages of an acute process, and two of a chronic character are severally to be traced to some form of blood contamination— fevers, struma, syphilis, alcohol, or the agency of cold and wet; and thus far they may be considered as varieties of the same morbid process; modified, however, by circumstances, perhaps, chiefly of idiosyncrasy.

The two remaining forms, the amyloid or waxy kidney, and the atrophic gouty contracted kidney, have nothing in common, except that the structural changes in both originate in a material deposited by the blood-vessels, not only in the kidneys, but in other and remoter organs. Amyloid substance is not confined to the kidneys in cases in which it infiltrates those organs; and in like manner the urate of soda, the parent of the nodular gouty kidney, is found in abundance in the fibro-cartilaginous structures of the joints, and even elsewhere in those cases where it has brought about destruction of the renal organ.

The amyloid degeneration of the kidney consists in the infiltration of the organ by a distinct and special product not confined to the kidneys, diffused through other organs, and recognised by its reaction with a solution of iodine. In the atrophic gouty contracted kidney there is also a deposit from the blood in the fibrinous matrix, and tubular structures of an excrementitious material; the urate of soda, which ordinarily is excreted with the urine, but which, when either in excess of what the renal cells can excrete, or unaccompanied by those salines by which it is retained in solution, exhibits a remarkable affinity for fibrous structures. When deposited in the kidneys it leads to loss of substance and of secreting power; as in the fibro-cartilaginous tissues of the joints it leads to loss of structure and of motor power. In this respect, then, is seen at once the distinction between the atrophic gouty kidney and the contracted granular red kidney. This latter is frequently met with in gouty patients of broken-down constitution, and it is also found in individuals exposed to, or suffering from, the contamination of lead, in whom gouty symptoms of greater or less intensity have occurred. But then this contracted granular red kidney is as often found in those who have neither had gout nor suffered from lead or any other metallic impregnation: in patients in whom the renal degeneration cannot be traced to any well-marked or definite cause. The pre-existence of gout is, therefore, not essential to the development of this particular form of renal degeneration, although the concurrence of gout with lead contamination by deteriorating the qualities of the blood and tissues may hasten or facilitate the structural changes going on in the kidneys.

The granular red kidney is, therefore, not essentially a gouty kidney. On the other hand, from my point of view the atrophic, contracted, red nodular kidney is in its very essence gouty. It is never associated with any other morbid condition. Microscopically its structure is identical with what is seen in the remains of kidneys that have been the seat of calculous nephritis or pyelitis. The cortical substance is then seen condensed and the tubes flattened and most of them obliterated, and such as can be traced denuded of glandular epithelium. The origin of these diseases or forms of disease is a deposit within the organ of uric acid, urates, or oxalate of lime, around which by slow accretion a calculus forms which works its passage towards the outlets of the kidney. If it be permanently retained within these or within the organ itself the renal structures eventually become atrophied and condensed; and the reason why contraction does not follow is, no doubt, from the permanently dilated or distended state of the organ from the retained urino-purulent fluid. But there is an example in the Westminster Hospital Museum of a kidney which was the seat of calculous pyelitis in which, from the escape of the calculus into the bladder, the cavity of the kidney was emptied of the purulent fluid, and the once dilated pelvis and kidney ultimately contracted, shrivelled, and became dense, firm, and nodular, even tubercular on its surface. The microscopic appearances of this kidney were similar to those of the atrophic nodular kidney in which urate of soda had been deposited in the elementary tissue of the organ.

CHAPTER V.

Chronic Albuminuria.

Symptoms, Diagnosis, and Treatment.

The forms of renal disease brought together in this group have many points in common both as regards deviations of structure as well as of symptoms, although in studying their origin they may be traced to very different causes. In each there is alteration of structure in the blood-vessels, thickening of the arterial walls in two, infiltration of amyloid substance in one, and rigidity of the walls of another by atheromatous deposit in the middle coat. In this group, however, the symptoms in the early stage are more obscure, less significant than in the former. There is albuminous urine in all. Purulent urine in none. Diagnostic casts, more or less typical, are found in the urine of each. The aspect of the patient is prominently characteristic in each. There is not the alabaster pallor of the acute form of albuminuria; but there is a sallow pallor,—a yellowishness of the skin, contrasting most strongly with the slight capillary injection on the cheeks, particularly of those previously addicted to habits of irregularity or intemperance. There is a pinched look in the features. The sprightly look of health and vivacity is gone, and the aspect, even to the least observant, is that of one of broken health. This is commonly present. It is worthy of observation, by way of contrast, that in more than one of the diseases comprised in the former group, in which extensive disorganisation of the kidney may happen, as in calculous nephritis or pyelitis, that the countenance or aspect of the patient throughout does not deviate from the ordinary expression of health, so that a patient may go about with a calculus in the kidney, passing purulent urine, and yet the countenance fail to indicate anything of the patient's malady, or convey any idea that such an important organ is the seat of such disease.

The fact is, that these, to a certain extent, may be regarded as local diseases; while in the group now under consideration the kidneys are but the exponents of a general and wide-spread defect running through the entire organism. The kidneys occupy this prominent position because the urine excreted by these disordered organs becomes from the first altered in quality and contains in its sediment evidence of the decay going on. The excretion of no other organ can be thus closely watched. The kidneys thus convey the earliest and the most certain intelligence of the approaching decay. There is less variety in the general symptoms than is observed in the acute forms.

Dropsy, as a symptom in the several forms of Bright's disease, may be expressed thus. In the acute form it is present from the first. In the chronic forms, in three, it appears in the course of their progress, not at the commencement; in one, the atrophied gouty kidney, it is either absent altogether, or exists only in a very trifling degree, and then only towards the termination. The amount or degree of dropsical effusion varies also with the form. It is most abundant and most diffused through all the internal organs, as well as the skin, in the acute form. It is next most abundant in the large granular kidney, particularly when associated with heart and lung disease; next in the amyloid variety. It is least in the granular, red, contracted kidney, and absent in the nodular or atrophied gouty kidney. The amount of albumen in the urine varies also with the form. It is most in the large, smooth, and the granular kidney, and in the amyloid form; it is intermediate in quantity in the contracted granular kidney, and least in the atrophic gouty kidney.

The commencement or the starting point of these several forms of chronic renal degeneration is with difficulty fixed. Unlike the acute form, which commences with febrile disturbance, the transition from an ordinary state of health to one of failing energy is imperceptible, and oftentimes unnoticed except through some accidental coincident. A slight cold, some bronchial catarrh, is marked by evidence of greater prostration than is at all commensurate with the local complaint. In professional life, or among those raised above the sphere of bodily labour, some mental strain or subject of business anxiety induces a degree of debility and exhaustion quite disproportionate to the event. In some, occasional embarrassment in the breathing, in others a sense of fatigue and painful weariness after the most moderate exertion, in others a peculiar and persistent headache, indicates the insidious advance of these chronic forms of degeneration and decay. There is one symptom which is common to all, and which is never absent—it is the total loss of appetite and inclination for food, and this without any corresponding disorder of the stomach.

These trifling and inexpressive symptoms are often passed by as insignificant; rest and tonics are thought sufficient for them. The state of the urine is often overlooked. It is perhaps not examined till some slight puffiness or œdema of the ankles and feet suggest the idea of dropsy and renal disorder. It may be instructive to trace the earlier symptoms in typical cases of each of these four varieties of chronic albuminuria disease, selecting them in the order in which they have been already placed according to their apparent cause or origin.

I. CHRONIC GRANULAR DISEASE; *Pre-existing blood poison —certain fevers, &c.*—The albuminuria or chronic renal disturbance, which in contradistinction to the acute disorder which a pre-existing blood-poison, such as scarlet fever, enteric fever, or erysipelas occasionally develops, declares itself very obscurely. Œdema is not at first present. There is little to be noticed beyond a general deterioration of the health which is often attributed to a slow convalescence. In some cases the interval between the subsidence of the fever and the symptoms of albuminuria is marked by complete convalescence and a return to the ordinary health; when headache, inappetency, and prostration without adequate cause may lead to a searching examination of all the organs, and then the urine is found albuminous. The result or sequel of such cases will depend chiefly on the constitution of the individual; in strumous children the disease may run into granular degeneration, and of course with a fatal termination. A careful examination of the sediment in the urine by the microscope will ordinarily determine this point. If isolated blood-corpuscles, or casts containing blood-corpuscles, indicate recent engorgement or hyperæmia of the kidneys, the albuminuria may be regarded as arising from an acute process, and the prognosis will in the main be favorable. If on the other hand the casts are simply granular with resplendent nuclei, or there be a disproportion of granular or fatty cells—imperfect or abortive epithelia—then the prospect of recovery is remote. In young people the headache is sometimes a prominent symptom, and may be followed by an epileptiform attack, which, however slight the convulsive paroxysm may be, is always to be regarded as a most unfavorable sign. If the urine contains evidence of recent engorgement of the kidney, the treatment should be in principle identical with that adopted in the acute disease.

If the diagnosis recognises the chronic form leading to granular degeneration, the remedial measures should include hot-air baths occasionally; a well-selected nutritious diet combined with appropriate preparations of steel; moderate exercise, avoiding all exposure to cold and wet, or inclement weather, and the body should be clothed in flannel.

In all these cases of chronic renal degeneration, particularly those traced to pre-existing blood or fever poison, the most hopeful road to relief is found in placing the patient in a pure and bracing atmosphere; either by the seaside, or in some elevated locality where the air is dry and invigorating. It is in such cases that a long sea voyage to a temperate climate has been found so beneficial.

In this class of cases, traced to whatever cause, the living in and breathing a pure exhilarating air is most essential.

Without this aid all other remedies are comparatively powerless. The part which the respiratory process plays in these disorders, or rather the part which the absorbed oxygen performs is, I think, not recognised sufficiently.

In these forms of chronic albuminuria the blood is impoverished; the blood-corpuscles deficient; the aspect tells it, analysis proves it. Effete material collects and accumulates, or is unoxidized in the tissues—and this not because there is any mechanical or pathological impediment to the entrance of air to the blood in the lungs, as might be the case in bronchitis, emphysema, and tubercle—but because a portion of the blood-corpuscles have become so deteriorated in their essential vital property of oxygen carriers, that material deposited from the blood remains unoxidized, or incapable of being broken up into products capable of excretion and removal.

Fatty and granular degeneration of important excretory gland-cells is the result; and this decay, or descent in the scale of vital energy, can only be compensated for, or removed, by a treatment which, while it aims through nutrition and steel to supply a more vigorous blood-forming power, will also, through the agency of an exhilarating atmosphere, enable the fresh-formed blood-corpuscles to assist in the process of oxidizing the (effete) fatty products of previous disease.

We know how large the capacity of the blood is for oxygen, for, while 100 volumes of water at ordinary temperatures and pressures will absorb only $2\cdot 97$ volumes of oxygen, the same quantity of blood will absorb from 17 to 20 volumes of the gas. Moreover, it appears that the blood-corpuscles possess very peculiar relations to ozone, absorbing it with considerable energy and parting with it with equal facility, acting apparently and pre-eminently as ozone carriers.*

From observations made on several cases of chronic albuminuria, which have been placed under circumstances favorable for breathing and inhaling air charged with ozone, I am disposed to place great reliance on pure air in the management and treatment of these cases.

* Carpenter's 'Physiology,' edited by H. Power, p. 175.

A sea voyage of some duration is often remarkable in its retarding agency to the onward progress of renal degeneration. A residence by the seaside, if accompanied by habits of early rising and exposure to the influence of the morning air, is often attended with the best results. The only care necessary is to avoid the damp chills of evening. The whole body should be clothed in flannel; if the patient's strength permit, he should endeavour to be out half or even a quarter of an hour before breakfast. After that meal, at least an hour, or even more if the condition of the patient permits, before noon should be in fine weather spent abroad. A carriage airing will suffice to those whose means permit, or whose bodily strength does not allow of walking. The afternoon may be spent in any way conducive to the pleasure and amusement of the patient; avoiding, however, all public assemblies, or gatherings of crowds in ill-ventilated rooms or theatres.

The mineral waters best adapted for these cases are those of a chalybeate character; and a residence of some weeks at Vichy, Pyrmont, Schwalback, or Spa, all of which have ferruginous springs, is oftentimes followed by very satisfactory results, and will be found most advantageous in those cases of chronic albuminuria traced to pre-existing blood poison.

It may be well to remark here, that it is in this class of case that albumen often continues so persistently in the urine. I have notes of several cases in which the urine contained a trace of albumen five years after the contamination of the system with the poison of diphtheria. The health in this long period had been uninterruptedly good. One gentleman continues the duties of an arduous profession without stint or reluctance. Another has married and has two, if not three, remarkably healthy scions.

The albumen in these cases of persistence usually undergoes a remarkable modification; its chemical reactions differ from ordinary albumen, and chemists apply to the modified albumen the term Albuminose.

Under the head of Albuminose, in the third part, this subject will be more fully discussed.

II. Chronic Granular Disease *associated with heart or lung disease.* — Valvular disease of the heart with more or less hypertrophy, chronic bronchitis, and emphysema,— these are the cardiac and pulmonary disorders which are the leading precursors of cardiac and pulmonary dropsy, with subsequent granular degeneration of the kidneys. It is not necessary for the present purpose to describe the cardiac or pulmonary symptoms which precede the appearance of dropsy. Suffice that a patient long, perhaps for years, a sufferer from chronic bronchitis, with probably more or less of emphysema, or who has long had a damaged heart—a systolic murmur, mitral or aortic, slowly shows symptoms of increasing debility. The ankles swell toward evening; the swelling disappears after a night's rest; week after week, with attacks of dyspnœa of increasing frequency and cough of more than usual severity, the anasarca of the lower limbs slowly rises higher, perhaps to the knees. The urine becomes scanty, dark coloured, and deposits copiously reddish or pink lithates. It is examined for albumen, and none found, or only a trace. Progressively, however, as the dropsy increases, the urine, still scanty and deficient, becomes more albuminous. Casts may now be seen in the sediment. They are chiefly hyaline or slightly granular; a few modified epithelial cells may occur in the casts, but many are isolated. There is usually no change in the character of the sediment throughout. The dropsy may by remedial measures be temporarily reduced and some relief given. But, notwithstanding, the progress towards a fatal termination continues unchecked, and the patient usually succumbs to increasing difficulties in the respiration, pulmonary œdema, with serous infiltration into the air-passages, being for the most part the precursor of death. But chronic bronchitis, emphysema, and rheumatic endo-carditis, although the parent of some cases of cardiac dropsy with attendant renal degeneration, are by no means exclusively so. Many cases of dropsy commencing in embarrassment in the moving power of the heart are unconnected with any valvular defect. The organ becomes defective in power, flabby and fatty in structure, although increased in volume in the form of hypertrophy so constantly associated with renal degeneration. Probably the blood is the starting point in the series of pathological conditions which follow. The heart labours to drive an imperfect fluid through capillaries offering obstruction to its passage.* The heart increases in volume, but not in power, and eventually its unequal action permits passive congestive conditions, which in such organs as the kidneys do not suffer disturbance in the current of blood through them without giving signs of altered function and eventually altered structure. A large proportion of cases of this description owe their origin to the ill-timed use of alcoholic stimuli. I say ill-timed, for in many there has been no excessive use of stimulants. Publicans and those engaged in the spirit trade furnish many examples of this class of disease.

The Granular contracted Kidney.—In the same insidious manner commences that form of renal disease which usually terminates in the granular contracted kidney, the cirrhotic kidney of Harley. In some cases the disease is associated with gout, in others both with gout and lead poisoning, while in others there is neither gout, nor lead, nor intemperance, nor any adequate cause to explain the category of symptoms which slowly and progressively advance to a fatal termination.

The starting point cannot be fixed. The health slowly breaks, fatigue is easily induced, and the failure in physical power is often associated with chronic wandering pains, called rheumatic; dyspnœa occurs on trifling exertion; the ascent of a few steps causes trouble. There may be chronic cough, with a wheezing respiration, aggravated by the slightest catarrh. There is complete inappetency. There is frequency of micturition, with an abundant flow of urine, which is more or less albuminous; it is pale, of low specific gravity, rarely above 1012, often 1005—1008.

At first the amount of albumen is very small, often only a trace. This will vary very much as the disease advances, but the proportion of albumen is never very large. Headache of

* See 'Med.-Chir. Trans.,' vol. li, a paper by Dr. George Johnson.

a peculiar type often occurs. In many cases these head symptoms are the precursors, sometimes of partial and temporary paralysis, facial, or sometimes of convulsive attacks of an epileptiform character, dizziness of vision, opacity of the lens, and fatty aspects of the retina.

This class of case advances onwards with uncertain speed. In some the dropsy amounts to no more than œdema of the ankles; in others, there is not even that, but this will greatly depend on complication. With liver disease there may be ascites. Embarrassment in the action of the heart may occur, and the dropsy will then become more diffused. Otherwise the dropsical symptoms do not take a prominent place. Symptoms in reference to the nervous system are very characteristic. Apopleptic seizures sometimes occur, but hemiplegic attacks unconnected with cerebral hæmorrhage are not unfrequent; they are often transient, and very limited, may be only facial, may be the upper extremity only, the extent of motor and sensitive defect being but moderate.

As the latter end approaches, the nervous system presents oftentimes anomalous conditions, simulating various forms of brain disease.

Delirium, which may be mistaken for meningitis; coma, alternating with convulsive movements, suggesting cerebral hæmorrhage, and these symptoms may subside and reappear in the most conflicting manner.

The urine throughout this disease is noted for its abundance. The scanty proportion of albumen in some, the greater amount in others. The uric acid occasionally varies, and it may be associated with urates now and then; but this is also a varying symptom. But there is one unvarying quality of the urine throughout, and attending every case of this form. It is the deficiency of the urea. This excrementitious product is retained within the blood, and from its supposed decomposition in that fluid poisons the nervous centre, and develops the anomalous symptom of the so-called uræmic poisoning.*

* See in Part III, under the section Creatin, another interpretation of the theory of uræmia.

The sufferers from this form of renal degeneration are of no particular class or calling. It may be seen among the wealthiest as among the poorest. It may be earned by intemperance, or by irregularity of life; excesses of any kind may predispose to it. Among the working classes the absorption of lead into the system is a forerunner; and if gout, or the gouty condition of blood be grafted on that metallic absorption, the disease is the more likely to be strongly pronounced.

In some constitutions, particularly if given to a luxurious and inactive life, in whom gout has lurked, but not very loudly proclaimed itself, and in whom the health slowly breaks, this form of degeneration is not unfrequent.

But in other cases this variety of granular degeneration can be traced or assigned to no other cause than either a natural or premature decay.

The organism has reached the limits of its vital activity, and following the law pertaining to all organized structure, decay commences with the decreasing vital activity necessary for the continuance and maintenance of life; and as the signs of this diminishing activity differ in each individual as to the period of life at which they are manifested, so also is there a difference as to the organ or organs which first announce the commencement of this decline.

In many this degeneration is premature, from a variety of causes, hereditary in some, acquired in others. While in a few the signs of this decay declare themselves only with advancing years, and after a life of healthy activity the boundary line of three score years and ten may be reached before the indications of renal degeneration become apparent; it is, therefore, in such cases more consistent with an enlightened pathology to consider this degeneration as but the expression of a natural law, rather than that of a superinduced disease.

Treatment.—The treatment of these cases of chronic albuminuria from the causes assigned is purely palliative. Decay has insidiously set in, and it is very rare to see it arrested.

Its onward progress may be checked; it may be retarded, but it cannot be averted.

The power of the digestive organs must be carefully studied. Nutrition must be considered of paramount importance.

Ferruginous tonics are of great use. A few drops of the perchloride of iron taken in seltzer water makes a factitious chalybeate of great value.

The gouty condition when present is best treated by the citrate of lithia; all lowering remedies, as preparations of colchicum, should be avoided. If purgatives are needed the potash salts are preferable. If hepatic symptoms suggest mercurials, the sulphate of potash and rhubarb should be given in preference.

The observations previously made on the influence of change of air are equally applicable in these cases.

III. THE AMYLOID FORM.—*Symptoms.*—It not unfrequently happens, particularly in hospital practice, that the surgeon requests the opinion of the physician on the condition of a patient who is suffering from caries or necrosis, or perhaps some form of strumous or syphilitic ulceration, or other form of wasting disease, with the object of ascertaining if there be any lurking constitutional condition, tubercle or otherwise, which may forbid, or render more than usually hazardous, the operation he may contemplate for the relief of the patient.

To take a typical case. The urine is examined and is found moderately albuminous. It has a specific gravity of 1005, or a little above. It is clear and of pale lemon yellow. The sediment is scanty, but in it may be seen a few hyaline casts, in addition to others of the same character, but having dispersed through them a distinct collection of fat drops. The aspect of the patient is pale and anæmic, and there may be some superficial anasarca.

The patient says he passes a great deal of water, and has done so for a long time past, and that he requires to pass it very frequently. There has been a great amount of bodily debility. Such a case is typical of the amyloid form of renal disease,* and is quite unfit to run the risks of a capital operation.

In other cases the patient may not be at the time suffering from evidence of struma or syphilis, but there may be such in his history, or he may have served in a tropical climate and been the victim of dysentery, for which he has been invalided, and whose health has not been benefited by his return to England. He loses what little strength or power he may have recovered; his feet begin to swell; he passes urine in increased quantities; there is great thirst, and I have known diabetes suspected. The urine is examined. It is of low specific gravity, with a scanty sediment, in which the same type of cast is seen as has been just described. Other organs are the seat of this form of degeneration, and evidence of which is scarcely ever wanting in the liver, spleen, and the blood-vessels of the intestinal canal after death.

Treatment.—The treatment of these cases of amyloid disease may be summed up in a very few words. Nutrition and steel are the chief agents. To a certain extent some modification of this may be required by the special complications of any individual case. A strumous patient with evidence of tuberculous taint may be benefited by cod-liver oil and iron in the form of the phosphate and iodide. If a syphilitic condition taints the constitution the iodides may be preferred, particularly those of iron, with the iodide of potassium.

Dr. Grainger Stewart speaks confidently of the value of strychnia in the gastric and intestinal complications which manifest themselves in this form of disease. He advocates the employment of the liquor strychnia of the pharmacopœia in five or ten minim doses, and thinks it gives tone and vigour to the digestive function.

The diarrhœa is a symptom which forces itself most prominently on the attention of the physician, and often baffles every suggestion or remedy which may be advised. Astrin-

---

* This connection between antecedent exhausting diseases and amyloid degeneration has been successfully traced by Dr. Grainger Stewart, Dr. Wilks, and Dr. Dickinson.

gents, opiates, lead, are all given with but a temporary check to this exhausting drain. More is to be obtained by a strictly enforced diet. Confining the patient to a farinaceous form, in which rice, sago, arrowroot, with milk, are the chief ingredients, is followed by great abatement of the irritable state of the bowels.

IV. THE ATROPHIC, NODULAR CONTRACTED GOUTY KIDNEY.—The symptoms of this type of renal disorder differ in very trifling matters from those which are observed to be significant of the contracted cirrhotic kidney. It has been already observed of the latter that gout seems to be one of the predisposing causes of that form; but it is also seen that it is not exclusively so. It would seem that gout and lead and alcohol, separately or conjoined, may influence the nutrition or healthy condition of the blood, which seems to be the parent of that and many other forms of degeneration. But in the true gouty, contracted nodular kidney the starting point appears to my observation to be in the kidney or kidneys themselves. For it must be recollected that this atrophy often prevails in one kidney more than the other. They are very rarely equally affected; and in the history of those cases it is generally found that the patient has at some antecedent period suffered from gouty nephritis, sand or gravel. Often there has been at some time or other hæmaturia.

This kind of kidney is found only in those who have either suffered unmistakably from gout, with gouty concretions, or who have suffered from gravel, &c. The nature of the change of structure has been already described. The symptoms of this form have at first nothing at all specially significant. A gouty individual, one who has deposits in his hands or feet, and who has been the subject of hæmaturia, lithiasis, or gravel, perhaps three parts of his life, loses appetite, fails perceptibly in physical power, makes abundant urine, of low specific gravity, with deficiency of urea, but with excess of uric acid probably, and containing a small quantity of albumen, only sufficient, perhaps, to say that it is albuminous.

The sediment from this urine is very small, and consists

chiefly of mucous corpuscles, with either uric acid, or mostly with a few crystals of oxalate of lime, and with these may be seen some very delicate small-sized hyaline casts. They are rarely accompanied by any granules or cells. I have occasionally seen, however, the casts slightly granular, and in its darker portion a crystal or two of uric acid, or of oxalate of lime, or some rounded spherical bodies, looking like urate of soda. These conditions, when present, are diagnostic of this gouty kidney.

The health and power of those with such urine may slowly yield. It is remarkable how protracted such cases may be. If the patient be surrounded by home comforts, and can have rest and quietude, a very long time may pass without any visible change for the worse. A capricious appetite, some form of gouty indigestion, flatulence, heartburn, palpitation and flutterings in the action of the heart, may be among the occasional disorders noted.

Then there may be headache with giddiness, and now and then temporary confusion of thought or embarrassment of memory—possibly of speech. Transient hemiplegia may occur; and other symptoms referable to the nervous centre may from time to time suggest the prognosis of apoplexy, cerebral softening, or effusion into the ventricular cavities. And in this direction the fatal termination most usually approaches.

The arterial structures within the brain are in most of these cases the site of atheromatous opacity or rigidity, and apoplectic seizures or cerebral hæmorrhage are the usual sequel to this form of disease.

Treatment.—There is nothing special in the treatment of these cases. So far as the gouty constitution is involved the general principles which govern the treatment of that diathesis must be observed here. The frequent desire to micturate may be to some extent moderated by small doses of the benzoate of ammonia with the citrate of lithia, or half dram doses of the camphorated tincture of opium may be taken with advantage at bedtime, with bicarbonate of potash and lemon juice. It is unadvisable, if it were possible, to lessen the activity of the kidneys so far as the quantity of an aqueous urine passing through them is concerned, for this activity of function keeps the tubes washed out, and prevents the deposit within them of urates or oxalates.

The diet, and particularly the liquids taken, should be so chosen as to prevent the least tendency to the increase of uric acid and the urates. The secretions from the bowels should be watched. Warm purgatives may be given with advantage when torpor or sluggish action betoken some inactivity in the functions of the liver.

The kind of wine or stimulants, as well as the amount, must depend greatly on the habits of the individual. Hock is a very good wine for such patients.

RENAL CYSTS.—*Cystic degeneration; Hydro-nephrosis.*— Renal cysts are of frequent occurrence in kidneys which are otherwise healthy. They are frequent in extreme old age. They are frequent also in certain forms of renal disease, particularly in the two forms of granular degeneration, the fatty granular, as well as the contracted granular kidney; perhaps more frequently in the latter—certainly more numerously. They vary in size from a mere microscopic object, detected only by that aid, through sizes from the minutest point visible to the naked eye, to cavities as large as peas, or even larger. A distinction has been made between the so-called microscopic cysts and the larger cystic formations. The larger ones are more frequently associated with healthy structures, the smaller microscopical ones with granular degeneration. Their origin and mode of formation has long been a matter of doubt and inquiry. They are of great pathological interest. But in a clinical point of view they are of less significance, for they cannot be determined during life, as they give rise to no symptom by which their presence can be recognised. This remark does not apply to certain cystic developments of very large capacity which are sometimes met with in early life constituting the condition called hydro-nephrosis, and which is occasionally met with in adult life. These are, however, disordered conditions appended to the kidneys, and not always exclusively developed within the organ.

These microscopic cysts, as well as cystic tumours or cavities of larger dimensions existing in the parenchyma of the kidney, are not only not recognisable during life, but a great diversity of opinion exists as to the manner and source of their formation. Some pathologists have supposed them to be dilated Malpighian corpuscles, others that they are dilated uriniferous tubes. Mr. Simon conceived them to be formed out of the epithelial element, and that more or less of interstitial cystic development was the precursor or essential antecedent to the contracted kidney. He conceives the smaller cysts to be simple nucleated cells, of the same size, probably, as the ordinary epithelial gland-cells of the tubes. From these cells, however, they are distinguished by their definite outlines and the transparent fluid within them. He thinks the germs out of which they may be formed might equally belong to epithelial development; so that according to varying influences, healthy gland-cells might grow or these fluid-holding cysts.* When these cysts are abundant they seem to occupy the place of the tubes, and they are often so numerous as to give to the cortical portion of the gland, when a section is made, the appearance of a collection of minute fovea or cup-like cavities. Mr. Simon thinks that if this new development could be suddenly subtracted from the volume of the kidney, if the cysts should have their fluid contents removed by absorption, how great a shrinking or contraction of structure must follow. The kidney would be reduced to a third or less of its former dimensions, and he imagines that if the fluid contents of these cysts should be removed, the result would be the falling together of the textures into a dense and comparatively smaller compass. The cause of the shrinking or contraction of the kidney is, therefore, traced to the removal by absorption of the contents of these cysts and the contracted granular red kidney is but a form or stage of cystic degeneration. Mr. Simon doubted whether the interstitial effusion of

* 'Med.-Chir. Trans.,' vol. xxx, 1847.

lymph exercised any effect on the subsequent contraction of the kidney. These views, both of the origin and mode of development of renal cysts, as well as the ulterior changes produced in the organs in which they are formed, are not in accordance with mere recent researches.

It has already been stated that the contracted kidney is now generally believed to arise from the shrinking of the fibrinous material, which a chronic or subacute inflammatory process has poured out, and is, therefore, in no way connected with the removal or emptying of the fluid contents of these renal cysts.

Rokitansky's opinion of the origin and mode of formation of these renal cysts is somewhat different. He conceives them to be developed from an elementary granule, out of which the nucleus arises, and this nucleus eventually grows or dilates into the cyst. These cysts, he says, are filled with granulated nuclei, and occasionally with spherical or polyhedral cells. Dr. Bristowe has contributed some excellent observations on these microscopic cysts,* and a very minute account of their structure and a well-executed drawing of their appearance will be found in the volume quoted below. Upon the origin of these cysts he admits that he can throw but little light. He thinks it, however, certain that they are developed in some way or other in connection with the renal tubes.

These microscopic cysts are found in greatest abundance in atrophied kidneys. They are comparatively rare in those kidneys in which larger serous cysts exist. In the best marked specimens of the large cystic disease Dr. Bristowe says the microscopic cysts were not present. Wherever these cysts have been abundant the uriniferous tubes were seen atrophied, and either deficient or invisible; so that the microscopic cysts and tubes may be said to exist in an inverse proportion to one another. He thinks, moreover, that the development of these cysts is intimately connected with the atrophy and disappearance of the renal tubes, and it may be fairly assumed that these cysts are the result not the cause of

* 'Trans. Path. Soc.,' vol. ix, p. 309.

the atrophy. Still the question has yet to be solved how and why they are produced in connection with the wasting and disappearance of these uriniferous canals.

Congenital cystic disease.—This may be considered as a malformation, or at least a pathological morphosis.

In the 'Pathological Transactions' for 1848—49 there is a description by Dr. Lever of these exceptional conditions of development. Dr. Roberts' work 'On Urinary and Renal Diseases' may be consulted for further information on this subject.

PART III.

THE URINE—ITS CLINICAL SIGNIFICANCE.

PART III.

THE URINE,

ITS CLINICAL SIGNIFICANCE.

From a clinical point of view it is scarcely possible to overestimate the value of the indications which are to be obtained in many forms of disease by a skilful examination of the urine. There is no excretion which, if properly investigated, can tell so much or so truly. Morbid derangements in any of the chief centres of life, the nervous, the circulatory, or the nutritive, are attended by expressive alterations in some of the qualities of this important excretion. It must not be supposed that a simple qualitative examination of the urine only can lead up to an accurate diagnosis of the seat or nature of the disease. But let the symptoms be taken collectively, subjective and objective, and add to them the results of a qualitative or, if necessary, quantitative analysis of the urine, and a more accurate diagnosis will be obtained than can be expected without the aid of such an examination. Doubtless there are many disorders, functional as well as organic, in which the urine will afford no direct information. Nevertheless, its negative condition, that is, the absence of any indication of disorder in it, will be of importance in coming to a right conclusion, and especially in those cases in which our judgment is governed by the process of eliminating from consideration all those functions which offer negative or healthy conditions.

Thus, in a case of delirium tremens or other form of nervous disturbance, a natural or healthy state of the urine would lead to a very different opinion and prognosis to that which would be suggested if an excess of phosphate of lime were found in it.

The subject of the urine in its physiological and pathological relations has been so exhaustively treated of late years, both by foreign and native investigation, and particularly in this country by Dr. Parkes, Dr. Thudichum, Dr. Roberts, Dr. Beale, and others, that it might be supposed a work of supererogation to introduce the subject here. But the present work is addressed chiefly to students and young practitioners, with the single object of promoting their clinical knowledge of renal disease; and to omit a description of those qualities of the urine which are significant of renal disturbance would have deprived the work of a large portion of its proposed utility.

In the former part, under the head of the different renal disorders, the qualities of the urine have been stated in general terms; but there was no fitting place in that part of the work for a special description of the best and easiest methods, for clinical purposes, for examining and determining the qualities of the urine as a guide to diagnosis.

In passing in review the altered conditions of this excretion, I propose to limit myself to those chiefly which are significant of renal disease; and I venture to think that, for practical clinical purposes, the following arrangement will be found most useful. It cannot be too forcibly impressed on the student and young practitioner that it is not from one quality or from the presence of a single ingredient foreign to its healthy state, that a correct judgment is formed. It is by comparing the deviation found with the condition of the patient coincident therewith which constitutes the basis of our diagnosis.

Thus the presence of albumen in any given example of urine tells but little as an isolated fact. But taken with the circumstances under which it makes its appearance, and in conjunction with some other qualities of the urine with which it may be associated, it then becomes most significant, and leads to a correct estimate of the disease. Thus albumen may be associated with blood-corpuscles, or with pus-corpuscles, or it may be quite independent of either. Whenever blood is present, or whenever pus is present, albumen will be present also. But blood may be significant of a great variety of disorders, both of the kidneys as well as of the urinary outlets; pus may also be the index of both renal, vesical, and urethral disorder, and albumen existing alone may also indicate renal disorders of a different pathological type.

To relieve the student from some of the difficulties which he may find on the threshold of this important branch of clinical training, I propose to place in succession all those properties of the urine, physical, chemical, or morphological, which are significant of renal disease. Only those chemical processes will be described which are capable of being practised on the library table of the practitioner. The physical qualities will be taken first, as the colour, odour, specific gravity, and quantity. The chemical properties, or those which are determined by some simple chemical process, will be placed next in order; and, lastly, the morphological or urines containing organic forms demonstrated by the microscope.

By reference to such a systematic arrangement of the several properties of the urine which are significant or otherwise of renal disease, the young practitioner will acquire greater facilities in diagnosis in these renal disorders, and will learn to place the indications of the urine in their true relation to any given form of disease.

THE URINE.

PROPERTIES OF THE URINE, PHYSICAL, CHEMICAL, AND MORPHOLOGICAL, SIGNIFICANT OR OTHERWISE OF RENAL DISEASE, AND OF CLINICAL IMPORTANCE.

PHYSICAL PROPERTIES.—*Colour—Odour—Fœtid—Ammoniacal or Aromatic—Clearness or Turbidity—Specific Gravity—Quantity—Frequency of Micturition.*

PHYSICAL PROPERTY.

Colour. The range of colour in healthy urine extends from pale straw, or lemon yellow, through various degrees of yellow of the sienna tone to a tint as dark as burnt sienna or brown sherry.

After strong exercise, or any great increase of perspiration, the urine will be deficient in water, while the urea, and urates, and uræmatin, are increased, and the colour becomes deeper in consequence.

It might be presumed that every student would be acquainted with this range of colour of healthy urine; yet I have often found students quite incapable of giving such a description of the different tones of colour observed in healthy urine, and I accordingly repeat it here as a standard of comparison to start from.

Chemists are unanimous in their opinion that the special colour of the urine is derived from the colouring matter or hæmatine of the blood.

For this colouring matter of the urine the name of uræmatine is proposed by Dr. Thudichum. It corresponds to the urophæine of Heller; and urozanthin, uro-erythrine, and urrhodine, are terms im-

PHYSICAL
PROPERTY.

Colour. plying that the colouring matter is yellow, rosy red, or red from some pathological cause.

The depth and tone of colour varies much even in health, ranging from a pale lemon or amber yellow to a tint of burnt sienna tone. This depends solely on dilution or concentration of the urine by diluents, exercise, temperature, and food.

Many disordered states of the system are, however, marked by great variations in the colour of the urine, and for clinical purposes the following list of deviations of colour, with the disease with which that alteration of colour is connected, may be found useful. It must be remembered, however, the colour of the urine[*] will be governed either by the dilution or concentration of the uræmatine or by the presence of other colouring matters derived from other sources, as when either bile pigment is present or when the hæmatin of the blood is directly admixed with the urine.

Interesting as is this subject both in a chemical or pathological point of view, it is of subordinate importance in a clinical sense. The colour of the urine taken alone will tell as little as the specific gravity taken alone; but, as many diseases exhibit in the urine very characteristic qualities of colour, and to the experienced eye afford a certain clue to the probable nature of the disorder, and facilitate the necessary examination into the several disordered functions, a list of the various gradations and quality of colour with the disease of which it is the probable index is here inserted.

For further information on the subject of the colouring matter of the urine in health and disease

---

[*] Schunck has investigated this subject with great success. See 'Proceedings of the Royal Society,' &c.

PHYSICAL PROPERTY.	
Colour. —	the works of Dr. Parkes and of Dr. Thudichum may be consulted with advantage.
Nearly colourless.	In hysteria, spanæmia, chlorosis.
Very pale lemon, straw, or gamboge yellow.	Not inconsistent with health; in cold weather or after diluents; in infancy; in chronic granular degeneration; in the atrophic contracted gouty kidney; in amyloid of the kidney.
Lemon yellow to a raw sienna tint.	In diabetes.
Amber yellow ranging through raw sienna yellow to pale burnt sienna.	In health, the range depending on dilution or concentration by temperature, diluents, exercise or nitrogenous food.
Burnt sienna.	In fevers and inflammatory diseases generally, with more or less deposit of urates on cooling. The deep colour of the urine in ague, called by Simon Uro-erythrine, is due to the formation of purpurin, an oxidized product of uric acid, and hence the pink colour of the urates, &c.
Greenish yellow, dirty yellow, smoky.	Early stage of scarlatinal dropsy, and in some cases of acute morbus Brightii, from admixture of hæmatin in small quantity always indicative of renal disease; also in the recurring inflammatory congestions of chronic granular and amyloid disease.

PHYSICAL
PROPERTY.
Colour.

Olive yel-   In jaundice and organic disease of the liver from
low to dark presence of bile pigment and bile acids.
olive yel-
low or olive
brown.

Reddish or   In all diseases of the kidney in which hæmaturia
pinkish,   occurs; nephritis; acute morbus Brightii; early
or rosy   stage of calculus of the kidney; in gravel occasion-
yellow.   ally; in gouty nephritis; in the early stage of
tubercle of the kidney.

Deep burnt   In fevers of all kinds, especially rheumatic fever;
sienna, or   in all inflammatory diseases.
reddish
brown.

Red, light   In all hæmorrhages from the kidneys, injury to
red, pink-   the loins, penetrating wounds, scarlatinal dropsy,
ish red, to   acute morbus Brightii, in gouty nephritis, in calculus
brown or   of the kidney, early stage.
even pur-
plish red.

The latter deeper tones are especially present in the hæmaturia from cancer and in polypus of the bladder, colour proportioned to the quantity of blood present.

Other colouring matters are occasionally found—of these, indican, according to Dr. Parkes (p. 201), may be contained in healthy urine. Indican and uric acid are found together, and they appear to have some relation to each other.

I have more than once found indigo in albuminous urine. On boiling, the albumen is abun-

PHYSICAL
PROPERTY.

Colour. —	dantly coagulated; on adding a drop or two of nitric acid, instead of the slight reddening which is usual from the oxidation of the uric acid present, a deep blue colour is developed. Its true significance is not known. In the cases in which I have found it associated with albumen and morbus Brightii the result was fatal.
Brownish black (*Melanourine.*)	This colour is derived from hæmatin, and implies the presence of blood from the kidneys. It is said that tar and creosote taken internally will produce black urine.

Odour.

PHYSICAL PROPERTY.

Odour. The smell of the urine in health is very characteristic—it is like nothing else—urinous. But in disease other odours intervene and supplant the ordinary urinous smell. These may be specified as —aromatic, fœtid, ammoniacal.

Aromatic. This may occur in health from certain substances taken with the food—asparagus; turpentine applied to the skin imparts an odour of violets; copaiba and cubebs transmit their particular aroma to this excretion.

In diabetes the odour of the urine is highly aromatic, like sweet briar, or new hay, or ripe apples, or chloroform.

Fœtid. In certain dyspeptic patients the urine acquires a very peculiar odour, varying from that of raw meat to that of a dilute creosote smell. In others it may be similar to decaying fish. Butyric acid is probably at the root of it.

Ammoniacal. In all cases where the urine undergoes molecular or chemical change before leaving the bladder. In diseases of the bladder with ropy phosphatic urine, the urea being converted into carbonate of ammonia by the decomposing agency of the abundant mucus and epithelium mixed with the urine. In typhus fever, malignant smallpox, and in other low adynamic states of vitality. The nitrogenous product of the waste of the tissues is held together so loosely that it undergoes decomposition while in the bladder.

PHYSICAL PROPERTY.

Clearness or turbidity, with sediment. With regard to clearness nothing can be deduced from this quality. But as regards turbidity or opacity it is important to note if the urine be passed clear, and clouds or sends down a sediment on cooling, or whether it be turbid or milky on passing from the urethra. All urines containing urates, even in small excess, will cloud on cooling, as those are insoluble at temperatures below 60°. If the urine on passing from the urethra be either cloudy or milky, then pus is present, and may be derived either from the urethra, bladder, or kidneys. If from the urethra there will be little shreddy particles of pus-corpuscles aggregated together; if from the bladder the fluid will be turbid and ropy; if from the kidneys the fluid urine will be opaque, or a pale creamy colour, and set at rest soon sends down a copious sediment of pus-corpuscles, leaving a clear surface of demarcation between them and the supernatant urine, which becomes now clear. A delicate pink line sometimes rests on the layer of pus-corpuscles and indicates the presence of a minute quantity of blood. The supernatant urine is always albuminous. The subject of sediments in the urine will be more fully considered hereafter.

Specific gravity. The specific gravity of healthy urine may vary through a very wide range according to dilution or concentration by accidental circumstances—1005 to 1024. These are occasional only, and the specific gravity of one period of the day will compensate for that of another, and the average become that of health. It is when the daily average is continuously low or high that a pathological condition is indicated. Urine of a very low specific gravity (1002—1004) is present in some nervous disorders—hysteria, spanæmia, chlorosis. 1004—1008 occurs

PHYSICAL PROPERTY.

Specific gravity. in several forms of renal disease. In chronic albuminuria with the contracted (cirrhotic) kidney, constituting polyuria or hydruria; also in the atrophic nodular gouty kidney, as well as symptomatic of the amyloid kidney.

The specific gravity is high in all fevers and inflammatory disorders, particularly in acute rheumatic fever, in puerperal peritonitis, in pneumonia, hepatitis, &c., also in cardiac, pulmonary, and hepatic dropsies. It is highest in diabetes, when it may reach 1040—1050.

Quantity. An accurate estimation of the quantity of urine passed in the twenty-four hours is a matter of considerable difficulty. For physiological purposes, as well as in the study of particular diseases, different observers have obtained reliable results, and the works of Dr. Carpenter for the first object, and those of Dr. Parkes, Dr. Thudichum, and as regards albuminuria Dr. Dickinson may with advantage be consulted for the second. But for ordinary clinical purposes (except in hospital practice, and then not always to be relied on) in general practice an approximation to the average daily quantity can only be attempted.

Quantity is diminished in cases of high specific gravity, except in diabetes, when it is enormously increased.

Quantity is deficient in all acute diseases; in all fevers; in the early stages of all dropsies. It is increased in all cases of low specific gravity. In hysteria, in the red contracted kidney, in the atrophic, nodular, gouty kidney, and in amyloid disease. As before stated it is enormously increased in diabetes.

PHYSICAL PROPERTY.

Frequency of micturition.

In most renal diseases the call to make water more frequently than in health is a prominent symptom. But it is a symptom of little value taken alone, but when associated with albuminous, bloody, purulent, or saccharine urine it is most significant. It is often a very distressing symptom, as the demand to pass water may recur every quarter or half hour from an almost uninterrupted sympathetic irritation at the neck of the bladder.

It is a prominent symptom in all the acute inflammatory diseases of the kidney; in nephritis, from whatever cause, wounds or rupture from bodily violence. In the early stage of acute morbus Brightii, in chronic morbus Brightii, particularly in the red contracted kidney, in the atrophic gouty kidney, in latent gout, in gravel or lithiasis, in calculous disease of the kidney or bladder, in tubercle of the kidney, in cancer of the kidney, in diabetes, in stone in the bladder, in urethral inflammation, gonorrhœa, and prostatic disease.

In some individuals, otherwise healthy, an acescent state of the urine from excess of uric acid will induce frequency of micturition. Errors of diet, or certain wines, such as champagne, will bring this on. A few grains, ten to twenty of carbonate of potash, with the juice of a fresh lemon, taken at bedtime, quickly relieves this occasional symptom. In elderly people, of either sex, frequency of micturition is often complained of, and it frequently occurs quite independent of any renal or other disease.

Chemical Properties.

CHEMICAL PROPERTY.

Acidity. It would be inappropriate to discuss here the complex chemical question of the source of the acidity of the urine. It will be sufficient for the clinical object of this work to state that chemists are generally agreed that the cause of the acidity is not due to the presence of any single acid, but rather to the presence of the acid phosphates and sulphates of potash and soda; to uric and hippuric acids, as well as probably to the lactic as well as to the butyric. See Dr. Parkes on the 'Urine in Health.' Some chemists have included in the list the oxalic or its acid salts, but the oxalic cannot exist as a free acid in the presence of soluble lime salts. The degree of acidity of any given sample of urine can only be determined by a volumetrical process, by a standard solution of pure carbonate of soda.

For clinical purposes in ordinary practice this estimation is not needed. It is sufficient to ascertain the fact that the urine has an acid reaction on litmus paper.

Healthy urine should always show this acid reaction. The degree of acidity is not easily ascertained; nor is it, for clinical purposes, of much importance, as, when excessive, it is always accompanied by deposits of uric acid or oxalate of lime, which more immediately indicate an excess of acid in solution.

Neutrality. A urine having neither an acid nor an alkaline reaction is said to be neutral. This may arise not

CHEMICAL PROPERTY.

Neutrality. from a deficiency of any of the acids or acid salts which contribute to its acidity, but to the neutralising agency of some alkali taken in with the food. Some fruits will effect this. All the saline effervescing drinks, the lemon kali, as it is called in the shops. The effervescing granular citrates, now so extensively used, will produce a like result. In health, during the digestion of nitrogenous food, the urine becomes neutral, and regains its acidity as complete digestion is accomplished.

It is a curious physiological fact, which the student must keep in view, that the state of the urine in regard to acidity is inversely as that of the stomach. In the first stage of digestion the stomach is in the highest state of acidity, supplying freely the gastric acid. At this time the acidity of the urine is at a minimum, or, more frequently, neutral. As the acidity of the stomach decreases, that of the urine increases. Animal food will induce a more decided neutrality than vegetable. In a purely vegetable diet the urine never reaches a state of neutrality during the stage of primary digestion.

Albuminous urine is occasionally neutral, the alkaline salts accompanying the albumen being sufficient to neutralize the ordinary acidity.

Alkalinity. Alkaline urine is a sure index of some disease, except in those exceptional cases in which large quantities of alkaline drinks have been taken, or after purgative doses of the tartrate of soda, or the citro-tartrate of potash and soda.

The alcalescence of the urine may be brought about either by the agency of ammonia, the product of the decomposition of urea, or by a fixed alkali, the result of some error in the primary and secondary stages connected with some disorder of the general

CHEMICAL PROPERTY.

Alkalinity.

system. Dr. Bence Jones pointed out this distinction, which is of the utmost clinical importance.* For when the alcalescence arises from the volatile alkali, the decomposition of urea is caused by the action of some morbid secretion within the urinary passages, chiefly in the bladder itself; and the treatment must be directed to that portion of the renal or urinary organs from which the morbid ferment is derived.

On the other hand, if the alcalescence arises from a fixed alkali, attention must be directed to the state of the digestive and nervous systems more particularly, the functional error of the first being most frequently connected with, or dependent on, disease or disorder of the latter.

To distinguish between the two sources of alcalescence the colour of the blue stained litmus, or the brown-coloured turmeric, is fugitive on exposure to a slightly warmer temperature if ammonia be present, but remains permanently blue or brown if a fixed alkali be the cause. It is to be further noticed that in alcalescence, from either cause, the earthy phosphates are precipitated: the fixed alkali causing, however, the precipitation of the amorphous phosphate of lime, while, by ammonia, the phosphates of ammonia and magnesia are thrown down in conjunction with the phosphate of lime: and the triple phosphate, in beautiful prismatic crystals, is abundantly formed, and easily recognised by the microscope.

Alkalinity from decomposition of urea by volatile alkali occurs in cystitis from whatever cause: in calculus, gouty cystitis, prostatic disease, injuries to the spinal cord, in tubercle of the urinary organs, kidneys, ureters, and bladder.

* See Dr. Carpenter's 'Physiology,' edited by H. Power, 6th edition, p. 415.

CHEMICAL PROPERTY.

Alkalinity. — Alkalinity from fixed alkali occurs chiefly in inflammatory disorders of the brain.*

In delirium tremens the presence of alkaline urine would justify the inference that organic mischief of an inflammatory character was added to the functional disturbance caused by an alcoholised blood.

Earthy Phosphates. Phosphate of Lime. — The general conclusion deduced from a series of observations made by Dr. Bence Jones† is, that in acute inflammation of the substance of the brain there is an excessive amount of phosphates in the urine; while in acute or chronic disease of the membranes there is no increase in the total amount of the earthy or alkaline phosphates.

In delirium tremens there is a deficiency of phosphates if no food be taken. In mania, melancholia, and even in the general paralysis of the insane, there were no marked alterations in the properties.

Chylous urine. — In some rare and exceptional forms of digestive error the urine passed has a somewhat turbid cloudy appearance, and on being set at rest and examined by the microscope, an abundance of oil or fat-globules are seen, free and unaccompanied by casts or cells. The urine coagulates slightly by heat and nitric acid.

‡ *Kystein.* — The urine of pregnant women occasionally presents a disturbed appearance, being cloudy on passing, and forming, when set at rest, a film on the surface as well as a subsequent sediment. The film on the surface soon exhibits glistening particles which the microscope proves to be crystals of the ammonio-magnesian phosphate mixed with highly resplendent granules of free fat. Lehmann considers this substance to consist of butyric acid and fat, phosphate

---

* Dr. Bence Jones. † 'Lancet,' March, 1850.
‡ Κύησις, pregnancy.

CHEMICAL PROPERTY.

Chylous urine. of magnesia and casein. Kystein was formerly thought a test of pregnancy. It is no longer considered so, as it has been found in the urine of women not pregnant.

Uric Acid.

Uric acid. The correct quantitative estimation of uric acid is a very difficult process, and requires skilful chemical manipulation, and is unnecessary for ordinary clinical purposes, and out of the question as involving too much time for the general practitioner. An approximation may, however, be readily attained by precipitation of the acid by hydrochloric or acetic acids in a tall conical glass. The crystals soon form and adhere to the sides of the glass, and their quantity may be rudely estimated by comparison. For clinical purposes, however, this process is not needed; for whenever uric acid is in excess and constitutes evidence of disorder the crystals are freely deposited or urates are formed in excess.

On the cooling of the urine uric acid, when in excess, will be seen in one or other of its numerous crystalline forms—as cubes or quadrants, rhombs or lozenges; hastate, stellar or columnar, or even barrel-shaped crystals, or in the form of crystals of oxalate of lime. Occasionally the uric acid is passed suspended in the urine as red sand, like cayenne pepper grains, and settles immediately as a copious deposit of red crystals. Or the uric acid has combined either with ammonia as a base, or with soda, forming the constantly occurring deposit of urates settling as soon as the urine has lost a few degrees of the temperature of the body. These urates are easily recognised by their complete solubility at a temperature above 65.

CHEMICAL PROPERTY.

Uric acid. Uric acid, as crystalline grains of red sand, is not significant of renal disease. It is the result either of the metamorphosis of nitrogenous food in excess of the requirements of nutrition, or it may be derived from the waste of the tissues, defective oxidation not having converted it into the more soluble and more easily excreted urea. Uncombined with any base it occurs after certain articles of food; some fruits will induce it.

It is not uncommon in childhood, or as the organism is advancing towards puberty.

In disease it is the accompaniment often of the gouty constitution, alternating with attacks of gout. It is the agent in lithiasis and gravel. It is the precursor often of renal calculus.

In combination with a base it is always in excess in the early stage of all fevers and inflammatory diseases constituting the lateritious or brick-dust sediment of early writers.

Urates. Simple concentration of the urine or deficiency of a due proportion of water will always cause the deposit of urates, even though uric acid or the urates may not be in excess of their proper proportion. All urine, therefore, of high specific gravity (except diabetic urine) will exhibit urates on cooling. In some diseases these urates acquire from the uræmatin particular tints. Sometimes they are pale, almost colourless, or at most cream-coloured. They may be stone coloured, or fawn coloured, or pinkish, or red (the brick-dust), or they may be of a delicate carmine. The several tints approaching the red are modifications of purpurine—the product of an oxidation probably both of the uræmatin as well as of the uric acid.

These pink urates are common in all inflammatory

CHEMICAL
PROPERTY.

Urates. disorders, but reach their highest development of rosy carmine in diseases of the liver, either inflammatory, simply congestive, or organic.

OXALURIA.

Oxaluria, the so-called oxalic diathesis. By far the most important associate with uric acid and urates is a crystalline product formed and deposited *after* the urine has been secreted in the kidneys, and for the most part after the urine has left the bladder and been exposed for a short space to the air. This product, easily recognised by the numerous and beautiful octohedral crystals which form so abundantly in some urines, and which chemically is known as oxalate of lime, has given rise to a peculiar hypothesis both as to its mode of formation and origin as well as to its significance as a symptom of disease. The subject is of so much importance and interest that I propose to devote a few pages to its illustration.

Oxaluria or the oxalic acid diathesis was first recognised as a distinct disorder or pathological condition of the urine by the late Dr. Prout,* who described the symptoms of this diathesis as belonging to the irritable or nervous class rather than to the congestive or inflammatory. He described this diathesis as marked by irregularity of the heart's action, intermission of the pulse, palpitation, flatulent disorder of the stomach, and more or less hypochondriasis, that is, depression and low spirits. Symptoms very typical of the gouty diathesis and very characteristic of the dyspepsia of that habit of body.

Dr. Golding Bird endorsed these views, and added by his popular work 'On Urinary Deposits' still further to the belief and the recognition of this

---

* 'Stomach and Renal Diseases,' p. 62, *et seq.*

CHEMICAL PROPERTY.

Oxaluria. diathesis under the term first employed by him of oxaluria. There were, however, not wanting those who, with equal advantage for observation and equal skill as chemists, soon expressed themselves as sceptical of the existence of this so-called oxalic acid diathesis. If not among the first to entertain, certainly the first to publish, in this country, his doubts of this oxaluric theory, was Dr. Owen Rees. In his work, published in 1845, 'On the Analysis of Blood and Urine in Health and Disease,' at page 147, he thus expresses his views:—"The state of the system on which the secretion of urine characterised by deposits of oxalate of lime depends is not well investigated. There appears some degree of probability that it is connected with the formation of lithic acid in excess, and with a state of system in which a considerable quantity of urea is secreted." Still later, in the Croonian Lectures for 1856, delivered before the President and Fellows of the Royal College of Physicians, he affirmed that oxaluria was not indicative of a diathesis, but that the oxalate of lime was formed after the urine had been secreted by the kidneys, and was derived from the uric acid and urates of that secretion. It is singular how long an erroneous theory will prevail when it originates with observers of reputation, and has been endorsed by other writers who follow and adopt a theory without sufficient proof of the correctness of the doctrine they propagate. There are many who still view this oxaluria as a distinct disorder, and conceive it to be something entirely different from the lithic acid diathesis, and who, continuing to accept the views of Dr. Golding Bird, believe the irritable form of dyspepsia with nervous depression, already mentioned, are symptoms peculiar to the presence of oxalate of lime rather than common

CHEMICAL PROPERTY.	
Oxaluria.	to the gouty habit with its marked excess of uric acid and urates, which, as shall be presently shown, are the chief sources and origin of the oxalic acid and oxalates found in the urine. I have adopted and taught these views for many years.* Let us see on what chemical and clinical facts these conclusions rest.

In endeavouring to trace the source of oxalic acid in the urine it must not be forgotten how constantly it is found as the product of the oxidation both of animal and vegetable substance.

The ethylene series of organic compounds (primary nucleus, C^4H^4, olefiant gas) (secondary nucleus, $C^4H^2O^2$, oxalic acid) has by Leopold Gmelin† and other chemists been investigated with a most exhaustive research. Many of these investigations are calculated to throw light on the source of oxalic acid in the urine.

Lehmann‡ observed that perfectly fresh urine presented no trace of oxalate-of-lime crystals, but that, allowed to stand exposed to the air, a great many of these crystals were found in the sediment. No one much engaged in the microscopic examination of urinary sediment but will confirm this observation.

Scherer thought the formation of the oxalate due to a kind of acid urinary fermentation. Lehmann, in a subsequent edition (1853), remarks on this theory of an acid urinary fermentation—"we must not forget that oxalate of lime may possibly be formed during this process. We know that there is a close connexion between the excretion of uric acid and the formation of this salt from the circumstance that in most specimens of urine both sedimentary

---

* 'On Dropsy,' 3rd edition, p. 333.

† 'Handbook of Chemistry,' vol. ix, Cavendish Society edition.

‡ 'Physiological Chemistry,' vol. i, p. 44.

CHEMICAL PROPERTY.

Oxa-luria.

and non-sedimentary oxalate of lime cannot be recognised by the microscope so long as the fluid is fresh, but as soon as crystals of uric acid present themselves crystals of oxalate of lime may be discovered."* And he further remarks—"Since uric acid when acted upon by certain oxidizing agents may be decomposed into urea, allantoin, and oxalic acid, we may assume that a portion of the uric acid may be decomposed during this acid urinary fermentation, and that oxalic acid is formed from it."

The chemical relations of oxalic acid to uric acid and the urates have been investigated by chemists of the highest reputation, and all concur that oxalic acid is one of the products of the oxidation of these animal products.

Uric acid is converted into oxalic acid and urea by the addition of oxygen and water.

Uric acid mixed with a fermenting agent (yeast) and an alkali with an elevation of temperature is decomposed into oxalic acid and urea.

Uric acid heated with water and peroxide of lead is converted into urea, allantoin, oxalic acid, and carbonic acid (Wohler and Liebig).

Chlorine water forms with uric acid alloxantin, alloxan, parabanic acid, and oxalic acid.†

Uric acid oxidised by nitric acid yields a large quantity of parabanic acid, oxalic acid, and ammoniacal salts.‡

Examples might be multiplied to prove that the oxidation of uric acid always results in the formation of oxalic acid, urea, and other compounds. The urates undergo similar decompositions. It is well known that Peruvian guano, which is almost

---

* 'Physiological Chemistry,' vol. iii, Appendix, p. 453.
† Gmelin, 'Handbook of Chemistry,' vol. x, p. 460.
‡ Ibid., p. 462.

CHEMICAL PROPERTY.

Oxa-luria.

pure urate of ammonia, becomes, by oxidation in the hold of the ship during the voyage, decomposed* and largely converted into oxalate of ammonia.

From the chemical point of view, then, it cannot be doubted that oxalic acid and oxalate of ammonia can be derived from the decomposition of uric acid by oxidizing agency.

These facts strongly support the opinion expressed by Dr. Owen Rees that oxalate of lime is produced after the urine has been secreted by the kidneys, and is derived directly from the decomposition of uric acid and the urates.

A very valuable contribution to the chemistry of this subject has recently been made by Edmund Schunck, F.R.S., published in the 'Proceedings of the Royal Society,'† entitled "On Oxalurate of Ammonia in the Urine." The process for obtaining the oxalurate is given in detail. It consists in filtering the urine through charcoal, by which organic substances are absorbed and separated, in addition to a peculiar crystalline fatty acid, which the author had recently discovered and described. The charcoal is then treated with boiling alcohol, and the filtered liquid evaporated, which yielded a brown syrup, in which a quantity of yellow crystals formed on standing. Subsequent washings with cold alcohol, and evaporation, ultimately yielded, by concentration, a quantity of crystalline needles, consisting of oxalurate of ammonia. The chemical and physical properties of these crystals are then detailed, all of which proved the crystalline mass to be oxalurate of ammonia. Mr. Schunck fairly admits that these experiments by no means decide the question

* Gmelin, 'Handbook of Chemistry,' vol. ix, p. 110.
† 'Proceedings of the Royal Society,' vol. xii, No. 95.

CHEMICAL PROPERTY.

Oxaluria.

whether the oxaluric acid exists originally in a free or combined state. He admits that in these experiments it is quite possible for a quantity of ammonia to form by the decomposition of the urea to saturate the oxaluric acid present. He is, however, inclined to the opinion that the ammoniacal salt pre-existed in the urine.

The presence of oxaluric acid in the urine thus established presents, Mr. Schunck thinks, an easy solution of the formation of oxalate of lime, so often found in this secretion. Oxaluric acid is composed of oxalic acid and urea, minus the elements of water. Its composition is identical with oxamic acid ($C_4NH_3O_6$).

Acids, alkalies, even water at a certain temperature, severally decompose oxaluric acid and convert it into oxalic acid and urea.

Urine containing oxaluric acid will, on boiling or simple exposure to the air, yield oxalic acid and urea; and the oxalic acid, with its powerful affinity for lime, would, at the moment of formation, unite with or decompose the lime salts always present in the urine, and thus the insoluble crystals of oxalate of lime would be immediately formed and deposited.

The opinion so long ago expressed by Dr. Owen Rees, that the oxalate is derived from the oxidation of the uric acid and the water is thus clearly made out, although the process is somewhat differently interpreted.

Regarding the origin of the oxaluric acid, Mr. Schunck distinctly says that there can be little doubt that in the animal frame, as in the laboratory, it is formed by the oxidation of uric acid, which is its only known source. This conversion of oxaluric acid into oxalic acid and urea may take place in any

CHEMICAL PROPERTY.

Oxa-luria. part of the urinary apparatus after the urine has been once secreted, and hence an intelligible explanation is offered of the possible formation of oxalate of lime, whether in the renal tubes or other parts of the renal structure; yet this can happen only when within the body it meets some agent capable of effecting its decomposition into oxalic acid and urea. Any undue proportion of either the acid phosphates, or even the basic phosphates of soda, might effect this, as both acids and alkalies are capable of effecting its decomposition. Mr. Schunck very justly concludes that this oxalurate ($C_4N_2H_4O_8$) may be considered the vehicle appointed by Nature for getting rid of oxalic acid in the least injurious form. Were this acid excreted as such, it would, by combining with lime, produce serious results, which are prevented by its passing off in a soluble state of intimate union with urea.

Such are the leading and most characteristic chemical relations of oxalic acid to urea, uric acid, and the urates.

It may now, I think, be admitted that if oxalic acid is thus derived from uric acid, it supports Dr. Owen Rees in his opinion that this acid is never formed in, or excreted from, the blood, its only probable derivation being uric acid; it will occur only when and after that acid has been secreted; and consequently the urine must be its only known source.

It has already been remarked that perfectly fresh urine exhibits no crystals of oxalate of lime, while the same urine, after exposure to the air, will supply an abundant sediment. If such presumably healthy urine be divided into two portions, and from one of which all the uric acid is precipitated by acetic acid or hydrochloric acid, and then filtered, and the

CHEMICAL PROPERTY.

Oxaluria. other left for some hours exposed to the air, the latter will furnish evidence of octohedral crystals of oxalate of lime, while the former, from which the uric acid has been removed by precipitation and filtration, will not contain a trace.

All urines, whether derived from healthy individuals or those suffering from disease of the most varying character, are capable of forming oxalate of lime crystals. The clinical significance, then, of these crystals as a sediment, is not what either Dr. Prout or Dr. Golding Bird supposed. They are not significant of nervous depression or hypochondriasis, nor of dyspeptic disorder, any more than they are significant of emphysema, chronic bronchitis, phthisis, or even diabetes, in all of which disorders abundant sediment of the oxalate not unfrequently occurs.*

The sediment of oxalate of lime, examined by the microscope, is seen as octohedral crystals, or as the so-called dumb-bell crystals. The octohedral crystals are usually of two forms :—(1) The regular octohedron, (2) and what may be called the pyramidal octohedron, the long axis exceeding considerably the rectangular base. Occasionally the appearance on the field of the glass is that of a square outline, with a black square centre, the angles of the black portion of the figure being opposed to the sides of the outer or light square. This is merely due to transmitted light. The dumb-bell crystals were thought by Dr. Golding Bird to be oxalurate of lime. But this is not likely, for oxaluric acid, in combination with ammonia, as it probably exists in its formation out of uric acid, splits up into oxalic and urea in the presence of lime salts to form oxalate of lime.†

---

* See also Dr. Parkes 'On the Urine,' p. 221; and Dr. Roberts' Urinary and Renal Diseases,' p. 51.   † See *antè*.

CHEMICAL PROPERTY.	

Oxaluria. I think the chemical facts already stated, and the conclusions deducible therefrom, will tend materially to dissipate the chaos of opinion which Dr. Parkes* justly pronounced to exist on this subject. Oxalate of lime, as a sediment in the urine, is of no other significance than as an expression of a lithic acid diathesis. When in excess it implies excess of lithic acid. When seen in urine immediately on being passed, before such urine has time to cool, it signifies that the oxalurate of ammonia has undergone decomposition within the urinary or renal passages, and that the possibility of calculus is strongly suggestive.

In several cases of renal calculus in the early stage of frequently recurring hæmaturia, I have seen the crystals of oxalate of lime entangled in the minute flakes of fibrinous coagula, which have evidently formed within the urinary organs. It cannot be formed in the blood, nor can it exist in the blood, although Dr. Garrod thinks he has detected it in that fluid.

The symptoms of what was formerly called and recognised as oxaluria are now to be accepted as indicative of lithic acid in excess, and they may be summarised as consisting chiefly in more or less frequency of micturition, occasional sense of heat, or even scalding, along the urethral passage, dyspeptic symptoms of varying character, flatulent distress after food, irregular, perhaps intermitting, pulse, occasional palpitation, sluggish bowels, depression of spirits, restless and unrefreshing sleep, with its accompanying sense of weariness and languor. These symptoms, it is well known, are those very frequently met with in gouty subjects of indolent habits, or of dyspeptics, whose gastric distress chiefly arises from errors of diet, to which, perhaps,

* 'On the Urine,' p. 222.

CHEMICAL PROPERTY.

Oxaluria. may be added the mental strain of an anxious occupation, and the want of the exhilarating influence of air and exercise.

The symptoms described by Dr. Prout, Dr. Golding Bird, and others, under the title of oxaluria, therefore, belong to a class of disorders characterised by mal-assimilation, the evidence of which is drawn, not only from the disturbances in the nervous, circulating, and digestive organs, but from the presence in the urine of an excess of one of the most important excrementitious matters of the urine, lithic acid.

The causes of this disturbance are various. They may be traced chiefly to over tension of the mental faculties, to the anxieties of business, or of professional life, coupled, perhaps, with errors of diet and neglect in many minor matters of the general health.

The hypochondriasis (low spirits) is sometimes accompanied by singular delusions, and the direction most frequently taken is in that ' of the sexual function; a dread of impotency is the all-predominating idea.

It is from this class of patients that the victims are gathered by the scoundrels whose advertisements used to disgrace some of the daily journals, but which now, a higher tone prevailing, are excluded from their pages.

Oxaluria. Treatment. Medicinal treatment can do something, but diet and regimen can do more for the relief and cure of these cases. The former should consist of such alterative remedies as will relieve the sluggish and torpid bowels. Small doses of blue pill and rhubarb answer best. Some cases at first may require brisk purging. When the tongue and evacuations

CHEMICAL PROPERTY.	
Oxa-luria.	indicate a certain improvement from these remedies, the mineral acids may be prescribed with advantage. Of these the dilute hydrochloric acid in ten or fifteen minim doses, with the tincture of orange peel, taken immediately before food, will be found very serviceable. The citrate of potash, the bicarbonate of potash dissolved in water, and taken with lemon juice, will afford relief to any frequency of micturition which troubles the patient.

A few steadily observed rules of diet and regimen will do more, however, than all the drugs of the pharmacopœia.

The diet should be plain: it may be summed up in these words. Good English fare, neither a vegetable diet, nor an animal diet prevailing. The amount of animal food to be proportioned to the bodily exercise the patient may take. Of drinks, those which have a slight tendency to diuresis are best. Sound claret and seltzer water, or hock thus diluted. All the effervescing wines, all kinds of beer, except the best and most thoroughly fermented, should be avoided. Dr. Prout announced that the diet of these patients should be assimilated to that required for diabetes. It was not without some hesitation that I at first doubted the soundness of this doctrine, but the experience of many cases of this class of patients has convinced me that so far from that regimen being advantageous it was calculated to retard rather than advance recovery.

I have long observed and been convinced that even diet in these cases is subsidiary, or less necessary than certain rules in respect to what may be called in contradistinction to medicine, physical treatment or regimen. There is no reason, indeed it would be absurd to prescribe for such patients the training necessary for an athlete; yet the

CHEMICAL PROPERTY.
Oxaluria.

principles on which the muscular and bodily powers of an individual are brought out in the highest efficiency are applicable to these cases of the lithic acid diathesis. The disorder depends on the presence or accumulation of effete suboxidized matter, which in the moderate exercise of bodily activity is converted in the organism into easily eliminated excreta.

It must be borne in mind that this form of disorder is almost confined to men, and that it is most frequently met with between two or three and twenty and fifty, a period of life when bodily inactivity is most prone to be the parent of disorder.

The first and one of the most essential rules in this branch of treatment is early rising. This necessarily involves an early hour for going to bed. While under treatment the patient should rise at six in the summer, and seven in the winter; and immediately on rising a sponge, shower, or cold bath should be taken, with a free use of the flesh brush, and brisk rubbing of the entire surface with a coarse hempen bathing towel. The toilet finished breakfast should follow. If in business, time enough should be allowed for this meal to avoid hurried exercise immediately after it. If the patient be of no particular occupation there is particular need for some healthy out-of-door exercise, from two to four hours on foot; or if the patient can command it better on horseback. A light mid-day meal to those accustomed to it may be taken between one and two. Some pursuit in the open air, of a cheerful and animating character, may occupy the afternoon. The dinner hour should not be too late, the meal a moderate one. Retiring to bed early, the hypochondriac will find, after a day thus regulated, his spirits rise, his appetite improve, his sleep re-

CHEMICAL PROPERTY.

Oxa-luria. freshing, and his bodily functions more healthy. Pursuing such a course, with or without medicine, according as any error of function may continue present, the desponding patient will rapidly feel his depression subside, and in a few weeks, or even less, the pulse regains its regularity, the heart no longer palpitates, food is eaten and enjoyed without the distressing flatulence, and sense of oppression while gastric digestion is going on, and the patient ultimately learns that by a little management, directed by common sense, he has regained his health, and mental vigour and activity.

With professional men, and those in commercial or trading business, the suggestions here offered cannot, consistently with their pursuits, be entirely carried out. The regular daily exercise, the out-of-door employment for health sake, cannot be followed. To these, where they can command the time, or their occupation or employment permits their absence for a period of from two to four weeks, a change of air, or sojourn to an upland or marine district should be recommended.

The combined influence of a bracing air, with early rising, the daily use of the bath, and as much out-of-door exercise as the season or weather will permit, will very rapidly produce a marked change for the better.

Hippuric acid. The chief importance of hippuric acid as a constituent of human urine is in relation to the conversion of benzoic acid taken into the stomach into hippuric acid in the urine. It is a physiological rather than a pathological point. Chemists are still at variance whether hippuric acid is a constant constituent of the urine.* The compound tincture

* See Dr. Thudichum 'On the Urine.'

CHEMICAL PROPERTY.

Hippuric acid. of camphor (paregoric) contains benzoic acid. Balsam of tolu also. Benzoic acid or the benzoate of ammonia has been prescribed under the notion that uric acid is converted into the more soluble hippuric acid.

In the use, therefore, of these remedies hippuric acid would appear in the urine, where, if in abundance, it may be recognised by the addition of cold nitric or hydrochloric acids. The crystals occur in a stellar arrangement, and also in long needle-like prisms. Crystals of hippuric acid are colourless; those of uric acid are always stained more or less of a sienna orange colour. This acid is said to be present in the urine in cases of chorea. I have not been able to verify this observation.

Urea. The chemical processes for the quantitative estimation of urea in any given sample of urine are of importance only to the pathologist or physiologist engaged in some special inquiry. For clinical purposes they are not needed. They require a considerable amount of manipulative skill, and exhaust more time than the practitioner can afford, or is commensurate with the value of the information gained for the treatment of any individual case.

Urea in health represents the metamorphosis, not only of the nitrogenised food received into the system, but also that of the nitrogenised textures, muscular and others, which are constantly in a state of transition during the waking activity of the body; for more urea is formed when awake than when asleep (Thudichum), more during bodily exertion than during a period of rest or inactivity. In fevers and inflammatory disorders more urea appears in the urine than in health. As in these disorders little or no food is taken from disinclina-

CHEMICAL PROPERTY.	
Urea.	tion, the increase in the urea can only be derived from the waste and disintegration of the tissues. Hence the exhaustion and attenuation after protracted fevers; as the febrile action abates, the amount of urea becomes less than the average. But as convalescence proceeds, and the restorative power of food begins to attest the recovery of the patient, the urea returns to its healthy proportion.

In some diseases, both in acute and chronic morbus Brightii, the amount of urea sinks below the average standard, because of the defective excretory power of the kidneys. This defect arises from the fatty and granular degeneration of the epithelial gland-cells, the blocking up of the uriniferous tubes by effete cells, and in two forms of the chronic type, by the destruction and denudation of the tubes. Thus, the more immediate products of tissue—metamorphosis, creatin, and creatinine—are not transformed into urea, and they, as excreta, are retained in the blood, and eventually lead to the fatal group of symptoms termed uræmia.

Urea is in excess in the following diseases:— In all fevers, enteric, typhus, scarlet fever, smallpox, rheumatic fever; puerperal inflammation, peritonitis, and all inflammations; in some cases of disordered digestion (Dr. Fuller).*

Urea is deficient in all cases of morbus Brightii, both acute and chronic, in all cases where the specific gravity is continuously below the average. The average daily excretion of urea in health may be estimated at $3\frac{1}{4}$ grains per pound weight of the body (Roberts).

For clinical purposes the following are the best methods for an approximate determination of urea in quantity:

* "On Excess of Urea in the Urine," 'Med.-Chir. Trans,' vol. li.

CHEMICAL PROPERTY.

Urea. Two or three drops of urine on a slide, left to spontaneous evaporation, will, when urea is in excess, develop some silky and prismatic crystals within a few minutes. Some urine poured into a watch-glass to which nitric acid is cautiously added, so that the urine will occupy the upper layer, will develop crystals of the nitrate of urea when the latter is in excess; or a test tube may be used, the tube having about one third of its capacity filled with urine, strong nitric acid is added and the tube plunged into cold water. If urea is in excess, crystals of the nitrate will soon form, and their characteristic form may be examined by the microscope.

The quantitative determination of urea requires all the resources of a chemical laboratory, combined with experience in the processes of volumetric analysis. To those who have time and inclination for these researches the work of Dr. Thudichum is particularly recommended, as well as the New Sydenham Society's edition and translation of Neubauer and Vogel's 'Qualitative and Quantitative Analysis of the Urine,'* where the test solutions and apparatus for volumetric analysis of the urine are fully described. The student should, however, know the principles on which these volumetric analyses are conducted, even though he never practise them himself.

To estimate the quantity of urea accurately, the whole urine of the twenty-four hours must be collected. If albumen be present, it must be removed by coagulation, by dilute acetic acid and boiling, and subsequent filtration. The phosphates must be removed by a solution of caustic baryta and

---

* A French translation of this work, by Dr. L. Gantier, has just been published in Paris. 1870. F. Savy. With coloured plates of the crystalline sediments, and a coloured diagram of the spectrum analysis of the urine.

CHEMICAL
PROPERTY.

Urea. nitrate and filtration, and a definite quantity is then measured off for volumetric analysis. The exact volume of baryta solution must be noted, so that the proportion of urine may be known from which to calculate the ratio of the urea.

Liebig first made known the chemical fact that urea unites with mercury to form a definite compound insoluble in water. Its composition is $\overline{U} + 4HgO$.

If nitrate of mercury be added to a fluid containing urea from which the phosphates and sulphates have been removed, an insoluble yellow precipitate of urea and mercury is formed from which may be calculated the amount of urea in combination. To obtain this result the more readily and accurately the use of standard solutions of the tests employed are necessary. These are prepared according to the French scale, and the measurement of small quantities are made in centimetric cubes. The solution of nitrate of mercury is made of known strength, so that one centimetric cube indicates ·01 gramme of urea.

The mercuric nitrate is introduced into a burette properly graduated in cubic centimetres. The test is cautiously and gradually added to the urine, the mixture being well stirred until no further precipitation is apparent. To determine the point of neutrality with accuracy is the essence of the process. A drop of a solution of carbonate of soda is added to a drop taken by a glass rod from the mixture on a white slab of porcelain or glass laid on white paper. So long as the mixture remains white the urine requires more nitrate; but when a drop on the slab exhibits a distinct yellow, the process is completed. The quantity of urea is then calculated from the number of cubic centimetres of the nitrate

CHEMICAL PROPERTY.

Urea. of mercury required to precipitate it, every centrimetric cube being equal to $\frac{1}{100}$th of a gramme of urea.

It is quite clear that except for the purpose of some special physiological or pathological inquiry this quantitative determination of urea is not needed. It is occasionally performed in hospital clinical teaching for the purpose of demonstrating the process; but in general practice if the amount of urea passed be a point of importance to know, the practitioner should seek the aid of an analytic chemist unless he be personally skilled in these methods of analysis.

The phosphates. The part which the acid and alkaline phosphate of soda plays in the economy of the urine is a purely chemical and physiological one. No doubt that the excess or deficiency of these phosphates may become an element of disturbance, particularly in relation to the amount of uric acid held in solution. Clinically the determination of the amount of these salts is rarely attempted; the process is too tedious and not required in every day practice. But it is otherwise with the earthy phosphates. Their presence in excess in the urine is of the greatest clinical importance. Their presence is easily recognised both microscopically and chemically.

The earthy phosphates. These are two—the ammonio-magnesian phosphate and the phosphate of lime. There is probably a triple salt formed by the decomposition of the urine by the volatile alkàli, or phosphate of ammonia, magnesia and lime, which crystallises also in prisms similar to the ammonio-magnesian salt. The phosphate of lime when precipitated alone usually occurs in the amorphous form. It is precipitated from the urine by a fixed alkali.

CHEMICAL PROPERTY.

The earthy phosphates. Dr. Roberts has, however, pointed out that the phosphate of lime may occur in the crystalline form; and a beautiful object it is. It has been mistaken for one of the stellar forms of crystallization of uric acid, and, until the publication of Dr. Roberts' paper, I always considered these crystals as uric acid, and more than once I have been extremely puzzled to account for the appearance of them.

The distinction between these stellar crystals of uric acid and phosphate of lime consists in the fact that uric acid crystals are always more or less coloured orange or orange brown, while those of phosphate of lime are invariably colourless.

The prismatic triple phosphate is always the result of the decomposition of the urea in the urine, whether it takes place within the body or after the urine has been passed. The exciting cause of this is the presence of an undue proportion of the organic constituents of the urinary passages, either of mucus or pus derived from the bladder, ureters or pelvis of the kidneys. If from the bladder, then the decomposition takes place within that cavity and the urine is passed ropy, muco-purulent, and ammoniacal, and the prismatic crystals can be seen in abundance.

When the pus is derived from the pelvis of the kidney, or from the calyces, or even the substance of the organ, then the decomposition does not take place till after the urine has been voided, and which, although mixed with pus, is fluid and homogeneous, although milky or creamy in appearance, and not on passing ammoniacal. This triple phosphate plays an important part in the augmentation of the volume of a stone in the bladder. The outer layers of most calculi taken from the bladder consist of this earthy deposit, and many large calculi will exhibit alternate layers of this and uric acid. These deposits are

CHEMICAL PROPERTY.

The earthy phos- phates. derived solely from the decomposition of the urea and the adhesion of them subsequently to the calculus as to a foreign body; for the same result happens when catheters are retained for surgical purposes in the bladder, and when foreign bodies have passed into the bladder by introduction through the urethra, either intentionally by the surgeon, or by the unskilful act of the patient himself, or in girls from a morbid and unhealthy irritation.

Various substances have in this way been introduced within the bladder, and invariably, let the composition be what it may, they are found coated with a layer of these earthy phosphates, thick in proportion to the time the substance has been in the bladder.

The foreign body has been a source of irritation. The mucous membrane of the bladder has become the seat of irritative inflammation, and the muco-purulent exudation has set up definite molecular changes in the urine, resulting in the decomposition of the urea, and the precipitation of this ammonio-magnesian phosphate forming an incrustation on the foreign body.

The presence either of the amorphous or crystalline phosphate of lime is of far different significance. The former is thrown down from the urine in the presence only of a fixed alkali. In speaking of the alkalinity of the urine it was remarked that the alkaline conditions required for the precipitation of the amorphous phosphate of lime were usually derived either from some error or disturbance of the digestive or assimilative organs, or from what is of more clinical importance from the disintegration of nerve substance by inflammatory action. In all diseases, therefore, of the brain of an inflammatory

CHEMICAL
PROPERTY.

type, the amorphous phosphate of lime may be present and should be sought for.

Earthy phos-phates. Crystal-line phos-phate of lime.

The clinical significance of this crystalline state of the phosphate is not yet clearly made out.

In Dr. Roberts' communication to the 'British Medical Journal,' March, 1861, some diseases are mentioned which led him to think that it was an accompaniment of some fatal disorder. He mentions having seen this stellar phosphate in cancer of the pylorus, in phthisis, and even in chronic rheumatism. I can corroborate the fact of its alternating with uric acid as mentioned by Dr. Roberts.

The conditions necessary for its formation must be a urine either neutral or feebly acid, and when it alternates with uric acid this has been the case. In the last case in which I observed it a few months since, a young lad of fourteen had suffered one or two attacks of slight hæmaturia; and a day or two afterwards some gravel was stated to be seen in the chamber-pot.

The father, who also had suffered from gravel, feared his son might have the same complaint, and brought him up to town. The urine was perfectly bright and clear when passed, very feebly acid, clouded by heat, which was dissolved with effervescence of nitric acid.

A sample was set aside for microscopic examination. Twelve hours afterwards it was noticed that the natural haze of mucus, instead of falling to the bottom, occupied the upper third of the urine; several glistening crystals were seen in it, which, examined by the microscope, exhibited the stellar* uncoloured character described by Dr. Roberts, and

* The prevailing form was stellar. See Dr. Roberts' 'Urinary Disease,' p. 67, fig. 14.

CHEMICAL
PROPERTY.

The phosphates. were quickly dissolved in acetic and dilute hydrochloric acid.

Two days later another sample of urine being bright and clear, and having a strong acid reaction, threw down a copious sediment of orange red crystals of uric acid, but none of the phosphate.

The phosphates of the urine are either earthy or alkaline:—earthy, as phosphates of magnesia or lime, or in combination with ammonia as a triple salt, soluble only in an acid urine: or alkaline, as phosphates of soda, either as an acid or a basic salt, and soluble in all states of the urine.

The earthy phosphates may readily be precipitated by caustic ammonia, but an estimate of the earthy phosphates alone is never needed for clinical purposes. But an estimate of the total amount of phosphates, both earthy and alkaline, is occasionally of importance, particularly in diseases of the nervous centres, for there is a marked increase in the total amount of phosphoric acid and alkaline phosphates excreted in all those inflammatory diseases of the brain in which metamorphosis of nerve tissue is taking place. Dr. Bence Jones* first established this fact. If we desire, therefore, to estimate the amount of phosphoric acid excreted with the urine, we must obtain the total amount of both alkaline and earthy phosphates.

The earthy phosphates are first precipitated by ammonia; the next step is to convert the alkaline phosphates into insoluble magnesian phosphates. A solution of sulphate of magnesia has a small amount of hydrated oxide precipitated by addition of a few drops of caustic ammonia; this should be quickly redissolved by the addition of hydro-

* 'Med.-Chir. Trans.,' vol. xxx.

CHEMICAL PROPERTY.

The phosphates.

chloric acid, by which a small amount of chloride of ammonium is formed; the whole is then rendered again alkaline by a fresh addition of ammonia; this is then added to the urine with a little acetic acid, by which the soluble phosphates are precipitated as ammonio-magnesian phosphates, which, with the other phosphates first precipitated, may be collected on a filter, washed, dried, and weighed.

Another and more accurate method for determining the total amount of anhydrous phosphoric acid is by volumetric analysis with the nitrate of uranium, and a solution of acetate of soda with free acetic acid to insure the complete precipitation of the phosphate of uranium.

For the details of this process the Sydenham Society's edition of Neubauer and Vogel's work may be consulted.*

The chlorides, as chloride of sodium.

The chief, in fact the only point of clinical significance in reference to the presence or absence of the fixed chlorides of sodium from the urine is in parenchymatous inflammation of the lung—pneumonia; in which disease, at the very acme of the inflammatory condensation of the lung, the chloride of sodium altogether disappears from the urine and reappears as the process of resolution advances. This fact, first noticed by Dr. Redtenbacher, was made the subject of an interesting paper in the 'Medico-Chirurgical Transactions' of 1852,† by Dr. Lionel Beale, and has since been constantly recognised in clinical teaching in cases of pneumonia. The practitioner should always in these cases verify the fact for himself, particularly watching the period of the reappearance of the chloride as marking a favorable element in prognosis.

* Or the more recent French translation noticed at p. 202.
† Vol. xxxv, p. 325.

CHEMICAL PROPERTY.

The chlorides, as chloride of sodium.

The process for all ordinary purposes is a simple one. The urine to be examined should be first acidulated with a drop or two of pure nitric acid, and afterwards a few drops of a solution of nitrate of silver added. The chloride of silver thrown down, which is insoluble in nitric acid, is proof of the presence of the chlorides, and if no cloud be produced, it is concluded that the chlorides are absent.

For all ordinary purposes this rough process is sufficient. But Dr. Beale in his paper pointed out a fallacy which might constantly occur from the presence of hydrochlorate of ammonia which is frequently present in the urine, and which would give to the nitrate of silver the same precipitate as if a fixed chloride were present. He, therefore, evaporates a given quantity of urine in a water bath to dryness. The solid residue is incinerated by exposure to a dull red heat till thoroughly decarbonised. Whatever fixed salts were left were dissolved in distilled water acidulated with nitric acid. The nitrate of silver solution would thus prove the presence or absence of the fixed chloride of sodium. In the paper just quoted Dr. Beale further showed that the chloride of sodium absent from the urine in the acme of the inflammatory process was stored up in the hepatized lung and made its appearance in the sputa, for the sputa in pneumonia contain a larger proportion of chloride of sodium than healthy pulmonary mucus. Hence Dr. Beale concludes that the absence of chloride of sodium from the urine during the stage of hepatization depends on a species of attraction between the inflamed lung and the chloride, and that when resolution occurs this force of attraction ceases, and whatever portion of the salt is not excreted with the sputa is reabsorbed and appears in the urine in the usual way.

CHEMICAL PROPERTY.

The chlorides. — The process for determining the amount of chlorides by volumetric analysis requires a standard solution of nitrate of silver and a saturated solution of chromate of potash. In this process the albumen, if any present, must first be removed from the urine.

The sulphates. — But little clinical importance is attached to any variation of these salts. Their presence in the urine is easily shown by acidulating the urine with a few drops of perfectly pure hydrochloric acid and then adding a solution of the chloride of barium. The white precipitate is insoluble in all the acids. These salts are increased in chorea, also in inflammation of the brain; in which latter disorder they are accompanied by an increase of the phosphates.

Alcohol. — The presence of alcohol in the urine is of physiological rather than of pathological importance. It was formerly supposed that all the alcohol received into the organism by the mouth in whatever form, whether as spirit or wine or other fermented liquid, was completely oxidized in the system, and disappeared as carbonic acid and water, with some other product of its metamorphosis. It was subsequently affirmed that the greater part passed unchanged through the kidneys and appeared as alcohol in the urine. Dr. Anstie, in his work 'Stimulants and Narcotics,' first questioned this announcement, and proved that only a small fraction of the alcohol received by the mouth is excreted by the kidneys, and it is now admitted that Dr. Anstie's conclusions are right, and that only a minute portion of the alcohol taken is found in the urine.

Dr. Dupré has made many experiments in this direction, and has shown that alcohol by a refined chemical process can be detected in the urine even several hours after the stimulant had been taken,

CHEMICAL PROPERTY.

Alcohol. and he has also *quantitatively* proved by these experiments that the total proportion of alcohol eliminated by the kidneys is an excessively small fraction of what had been received through the mouth.

Dr. Dupré's process for determination of the presence of alcohol in the urine.—The urine, in as large a quantity as can be obtained, is carefully distilled, and the distillate re-distilled carefully, and, if necessary, even a third time. The distilled fluid is then subjected to the action of bichromate of potass and sulphuric acid, which oxidizes any alcohol present and converts it into acetic acid. This is subsequently recovered by distillation, and the fluid obtained saturated with baryta carbonate, and the acetate crystallised, dried, and weighed. The loss by weight after incineration gives the amount of acetic acid, which by calculation can be reduced to the amount of alcohol from which it had been produced. It is quite clear that this process can only be conducted by an experienced and expert chemist.

Another method simply qualitative has been proposed by M. Lieben, and of much easier application if it can be relied on. It is said by this new test that alcohol was detected in the urine of a man half an hour after he had drank a bottle of wine; and that fresh samples of urine taken an hour later, and one examined two hours and a half after the wine had been taken, supplied evidence by this test of the presence of alcohol.* The test is said to be capable of detecting one part of alcohol in 12,000 parts of water. This very delicate test depends on the conversion of alcohol into iodoform in the presence of iodine and an alkali. Iodoform crystallises in hexagonal plates or six-rayed stars.

* See 'Nature,' No. 2, p. 61, "New Test for Alcohol."

CHEMICAL PROPERTY.

Alcohol. The method of applying the test is to distil carefully the urine. The distilled liquid is then poured into a test-tube into which has been introduced a few grains of iodine and a few drops of potash solution, and the whole heated. If alcohol be present a yellow crystalline precipitate of iodoform occurs, sooner or later, according to the dilution of the liquid. The microscope demonstrates the crystalline forms above mentioned, which gives greater certainty to the crystalline character of the precipitate.

Albumen. In a pathological sense the presence of albumen in the urine occupies the very highest place of importance. Its connection with, and indeed its significance of one of the most fatal class of, disorders, entitles it to precedence before all other substances foreign to healthy urine. It may be derived from various sources or through various channels; but till its source and origin are made out, albumen in the urine will be the source of anxiety both to physician and patient. Whenever blood or pus is present in the urine, albumen will be there also, as it is a constituent of the serum of the blood as well as of the liquor puris. These, for the present, may be dismissed from further notice, as may the presence of a trace of albumen in the urine of men, containing gonorrhœal pus or gleety mucus, or in the urine of women suffering from leucorrhœa. It is when unaccompanied by blood or pus, and when it is more or less persistent or permanent in its appearance in the urine, that its presence assumes such pathological importance, and becomes indicative of serious renal disorder.

The following are the diseases in which albumen occurs in the urine, in the first group permanently, in the second temporarily :—

CHEMICAL PROPERTY.

Albumen.

GROUP I. *Permanent.*—Acute albuminuria, chronic albuminuria, in every form—significant of inflammatory, granular, fatty, amyloid, and atrophic degeneration of the kidneys.

In cases of cardiac or pulmonary disease, with dropsy, from obstructive circulation through the kidneys. In valvular diseases of the heart; in chronic bronchitis; emphysema with cardiac dilatations; in calculous and tubercular disease of the kidneys with purulent urine.

GROUP II. *Temporarily.*—In many blood poisons; in scarlet, enteric, and typhus fevers; in diphtheria; in erysipelas; in severe pneumonia; in cholera; in seminal emissions.

Associated with blood or pus—in nephritis; in all cases of hæmaturia; in gonorrhœa, gleet, and stricture; in leucorrhœa; from the action of turpentine,—of cantharides, after the ingestion of certain articles of diet, as in some cases of disordered digestion, after eating shell fish, crabs, or lobsters, &c. In purpura hæmorrhagica albumen may appear in the urine without blood.—Dr. Parkes.

Dr. Owen Rees has pointed out the fact that the resinous principle of copaiba gives to the urine a cloudiness on boiling, which may be, and has been, mistaken for albumen.

Processes for estimating.—Albumen in solution of an acid fluid will coagulate at any temperature above 140° Fah.; if alkaline no coagulation takes place. To determine its presence in urine, a test-tube is half filled with urine, and if not acid, to be acidulated by dilute acetic acid and boiled. The albumen, if present, clouds the fluid and forms, according to the proportion present, a small or abundant, a

CHEMICAL PROPERTY.	
Albumen.	fine or coarse, coagulation, which, when at rest, settles as a sediment. The height it occupies in the tube forms an approximate estimate of the amount of albumen in the tube. Thus, if we suppose the tube graduated into twelve parts, we estimate the sediment as $\frac{1}{12}, \frac{1}{8}, \frac{1}{4}, \frac{1}{2}$, &c.

This is sufficiently near for clinical purposes. It is advisable to add a drop or two of nitric acid after boiling, to ensure the solution of any earthy phosphate which may have formed by the disengagement of carbonic acid.

When there is but a trace of albumen present, it may the more readily be detected by half filling a test-tube with urine, and pouring down the side of the tube a small quantity of nitric acid, so carefully that it descends by the side of the glass by its own gravity, and occupies the bulb of the tube without mixing with the urine. At the line of junction between the acid and the urine a delicate cloud is formed if albumen be present. Uric acid in excess will sometimes occasion a doubtful cloud. Under the microscope a drop from the cloud will be granular if it be albumen, crystalline if uric acid.

A very minute quantity not perceptible by heat and nitric acid, or nitric acid alone, may be rendered apparent by tincture of galls, or by absolute alcohol or chloroform.

An approximate estimate of the amount of albumen in any given sample is all that is usually needed in clinical practice. To obtain an estimate of the total amount passed in the urine in the twenty-four hours is only required in certain special inquiries. The reader is referred to the works of Dr. Thudichum, Dr. Parkes, and others for further information on this point.

CHEMICAL PROPERTY.

Albu-men.
If it be required to determine more accurately the amount of albumen by weight in any given sample of urine, an accurate balance, a weighed filter before use, and a competent drying apparatus are necessary. The coagulated albuminous fluid is thrown on the filter, and subsequently washed frequently with hot distilled water. The filter and filtrate are then carefully dried and weighed; deducting the weight of the filter gives the weight of albumen.

Albumi-nose. Modified albumen.
It occasionally happens that in the favorable progress of cases of albuminuria a time arrives when the ordinary tests by heat and nitric acid separately fail to detect the presence of albumen. This is frequently the case in those who have had albuminuria after diphtheria. Sometimes nitric acid alone will not detect it, at others boiling will not detect it. Sometimes boiling produces a cloud, which is taken up by nitric acid, which would indicate the presence of phosphates, but on cooling a granular deposit of albumen falls. Sometimes nitric acid produces a cloud, which is taken up by boiling, and reappears again, as in the reverse process, on cooling.

This modified form of albumen is, however, always distinctly made manifest by tincture of galls. The cases in which this modification of albumen occasionally occur are those in which the assimilative process is not properly performed, so that errors of diet, and a certain form of imperfect digestion, are the most frequent causes. I have met with it more than once in patients after eating shell fish. I have met with it also in those of a nervous temperament, anxious about every symptom of their health, who examine their own urine, and, finding what they

CHEMICAL PROPERTY.

Albuminose.
Modified albumen.

conceive to be albumen of the blood, forthwith give themselves up to the fear of commencing Bright's disease. Healthy out-of-door exercise, the mineral acids, and care in the food they take, will soon cause the disappearance of this suspicious state of the urine. The most persistent case of this disorder occurred in the person of a clergyman who suffered from psoriasis, and in whose urine this substance had been detected by himself, as he was a dabbler in chemistry, taking the specific gravity and degree of acidity in every quantity of urine passed night and morning. I was at first disposed to think, as uric acid was present in abundance, that it was the cause of the apparent eccentric behaviour of the urine towards acetic and nitric acids, but it was not so; the sediment, however obtained, was always resolved by the microscope into the fine granular appearance characteristic of albumen. The tincture of galls, moreover, set the question at rest.

The properties and chemical composition of this form of modified albumen, the deutoxide of albumen, as Dr. Bence Jones proposes to term it, require further investigation.

Sugar.

Sugar or Glucose.—Sugar, it is said, has been found in minute quantity in the urine of persons otherwise healthy, and particularly after some varieties of food. This is probable, as in the so-called intermitting diabetes the urine of digestion is saccharine, while that after ten or twelve hours' fasting, that of the morning (*Urina sanguinis*), contains none. I once found sugar in the urine of a drayman, a large, bulky man, who drank several quarts of beer daily. Dr. Bence Jones found sugar in the urine of patients who had taken ether as an anæsthetic. Chloroform produces like effects. Thus sugar may appear in the urine without the patient

CHEMICAL PROPERTY.

Sugar. having diabetes. To constitute this disorder, not only does the urine contain a notable quantity of sugar, but the amount of urine passed greatly exceeds the average quantity of health; it is of very high specific gravity—1030 to 1050, perhaps higher—has an aromatic odour (not always), and is usually associated with thirst, unappeased hunger, and emaciation.

The detection of sugar in the urine in an ordinary case of diabetes is not difficult. Certain tests are necessary, and particular methods of manipulation required.

The usual copper test (Trommer's) depends on the property possessed by grape sugar, and other organic products, such as all the essential oils, of de-oxidizing many metallic oxides, or reducing them from a higher to a lower state of oxidation, or even to metallic purity. Those metals only are employed whose oxides of the lower grade possess characteristic colours.

The following are the most characteristic tests for sugar in the urine:

Liquor Potassæ.—The suspected urine is boiled in a test-tube with half its bulk of liquor potassæ. If sugar be present, the mixed fluids acquire a deep orange-brown colour from the formation of melassic acid, for it is one of the properties of sugar to be converted into this acid (called by some chemists the sacchaluric acid) when boiled with a caustic alkali. It is known as Moore's test.

The Copper Test, called Trommer's Test.—Tests required, liquor potassæ and a solution of sulphate of copper. The liquor potassæ must be pure and free from carbonate or sulphate. A large sized test-tube is filled to one third of its capacity with the urine; two drops of the solution of the sulphate of copper

CHEMICAL PROPERTY.

Sugar. should now be added, and the whole made strongly alkaline by adding the liquor potassæ. A greyish-green colour is formed, which is a hydrated oxide of copper, which is redissolved by the excess of alkali, and the solution becomes blue. The whole is now gently boiled over a spirit lamp. A bright orange colour is first produced if the quantity of sugar be large. As the heat increases this changes to a deep chocolate or coffee brown, or even red, when the suboxide is reduced to metallic copper. The chief precaution in using this test is to avoid using too much sulphate of copper.

These two tests are, practically, the most useful for clinical purposes, as the ingredients are always ready at hand, and the examination occupies but a few minutes.

Of the other tests based on the same principle as Trommer's, one is known as Hunefeld's, which consists in the reduction of chromic acid, the solution of which is an orange red, becoming of a brown bistre colour when added to urine containing sugar and exposed to strong light, first being slightly warmed.

This test is fallacious, as many of the organic constituents of the urine, uræmatine particularly, deoxidizes chromic acid.

The use of the blue hydrated oxide of copper, known as Capezzuoli's test, is similar in principle and application to Trommer's and not nearly so handy. Liquor potassæ is added to the suspected urine in a tall glass, to this some of the blue hydrated oxide is added, and the mixture set aside; in a few hours the edge of the deposit becomes yellow, and eventually red, from the reduction of the suboxide and metallic copper.

Runge's test by sulphuric acid, and various other reduction tests, such as the bichloride of mercury,

CHEMICAL PROPERTY.
Sugar.

the ammonio-nitrate of silver, or the bichloride of tin, are not applicable to clinical practice, and their description is omitted in consequence.

Fehling's solution.—This is a modification of Trommer's copper test. It is a solution of sulphate of copper in combination with a solution of tartrate of potash, and soda, and liquor potassæ. The mixture consists of 34·65 grammes of crystallised sulphate of copper, 173 grammes of tartrate of potash and soda, with 80 grammes of caustic potash in one litre of water. This mixture is readily decomposed by the action of light or carbonic acid, and exposure, therefore, to the air renders its indications fallacious. But this may be easily obviated by keeping the solution of sulphate of copper in one bottle and the solution of the tartrates and caustic potash in another, and mixing them in equal proportions when required for testing. Care must, however, be taken that, if for volumetric analysis, these proportions are accurately preserved.

Quantitative estimation.—In the treatment of diabetes it is of considerable importance to be able to estimate with a moderate degree of accuracy the amount of sugar per ounce voided by the patient.

To determine the absolute quantity passed in the twenty-four hours is not necessary for therapeutical purposes. It is a very tedious and uncertain process. The nearest approach of any practical value is to take two samples of the patient's urine; one passed shortly after food, and another passed after several hours' fasting. That passed in the morning on rising is the best. These two periods represent the amount of sugar probably derived from the food, the latter from the blood, and represents the products of the morbid gluco-genesis alone.

CHEMICAL PROPERTY.	
Sugar.	*Estimation of.*—There are three methods which may be employed for the determining the proportion of sugar per ounce. I. *By volumetric analysis.* Fehling's copper test just described is so prepared that a measured quantity corresponds to a known weight of grape sugar. But the process occupies too much time for clinical purposes, and moreover requires a good deal of manipulative skill, which every practitioner does not possess. The two others are simpler and even more reliable methods, as they require little or no special chemical skill.

II. *By fermentation.* The most elegant of these is that proposed by Dr. Roberts, of Manchester, and is based on the fact that every degree of density lost, as estimated by the urinometer, by the fermentation of saccharine urine, corresponds to one grain of sugar converted into alcohol, carbonic acid, and water, so that if before fermentation the specific gravity of the urine was 1040, and after fermentation was observed to be 1018, this would correspond to twenty-two grains of sugar to the ounce of urine which had disappeared by fermentation.

It is a process of very easy application, and requires about twenty-four hours for its completion. Having taken the specific gravity of the fresh passed urine, from four to six ounces are placed in any convenient glass vessel; a tall cylindrical glass is the best; to which must be added a morsel of German yeast, or ordinary brewer's yeast if the former is not to be obtained, and the vessel placed in some warm place or receptacle which will ensure the temperature not falling below 80° nor rising above 100° for the next twenty-four hours.

At the end of this time the urine will have become turbid, with a frothy scum on the top, and a sediment of the lees at the bottom.

CHEMICAL PROPERTY.

Sugar. The urine is now completely decanted, and if still warm, allowed to fall to the temperature of the air, and the specific gravity of the fermented urine taken. The difference in degrees before fermentation and after corresponds to as many grains of sugar per ounce converted into alcohol, carbonic acid, and water.

III. *By volume or weight of carbonic acid produced.* This consists in calculating the amount of sugar lost by fermentation from the *volume* or *weight* of carbonic acid evolved.

Thus, one cubic inch by volume of carbonic acid corresponds nearly to one grain of sugar; or, more accurately, 100 cubic inches by volume of carbonic acid equal 106·6 grains of diabetic sugar, while 100 grains by weight of carbonic acid equal 225 grains of diabetic sugar. To obtain these results considerable experimental skill is required, besides apparatus, which is only to be found in working laboratories. An accurate balance turning to the 0·001 of a grain; Liebig's bulb apparatus, with connecting tubes to dry the gas by passing it over fused chloride of calcium, and then absorbing it in the bulb filled with fused potash, all of which must be accurately weighed before and after the fermentation is finished. To estimate the sugar by the volume of carbonic acid, a mercurial trough, a graduated glass receiver to collect the gas over mercury with a properly adjusted receptacle for the urine, with securely adapted connecting tubes to conduct the gas to the graduated receiver, are required. It must be evident that these processes are not adapted for ordinary clinical practice.

Dr. Roberts' method is sufficiently accurate for all therapeutic purposes, and requires but ordinary care.

CHEMICAL PROPERTY.	
The polari-scope.	A very accurate estimate of the amount of sugar present in any given fluid may be obtained by this instrument. It is, however, not likely to be in the possession of those who are not working analytic chemists. Nevertheless, the student of medicine should be acquainted with its use. It is now very commonly employed in commerce, particularly in sugar colonies for testing cane juice, syrups, &c. The instrument must first be carefully adjusted, directions for which are usually given with the instrument. The urine to be examined must be free from all turbidity, and as colourless as possible; it must be filtered, therefore, through animal charcoal, any albumen present being first removed. The filtered urine is then carefully poured into the tube of the instrument, care being taken to exclude any air bubble, which is done by filling the tube above the brim, and dexterously sliding a glass cover sideways so as to bring the fluid accurately in contact before screwing on the cap. A strong light is necessary. The observer looks through the eye-piece at the light, and shortens or lengthens the tube, until a line dividing the field into two halves vertically is distinctly visible.

As the tube is turned from zero to 90°, the prismatic colours appear in the following order—yellow, green, blue, violet, red. Some practice is necessary to enable the eye to determine when the spectrum is divided into two equal halves of colour of equal intensity. When the two halves are similar in colour, the scale is read off, and the number of degrees from zero indicates the amount of sugar present.

In a very useful little work, 'A Guide to the Examination of the Urine,' by Dr. Wickham Legg,

CHEMICAL PROPERTY.

The polariscope. there are very good directions for the use of the polariscope for the estimation of sugar.

Cystin. Cystin is a substance of very rare occurrence in the urine. It was discovered by Dr. Wollaston in calculi, but it has been observed dissolved in the urine, and deposited after being passed as six-sided plates and hexagonal stellæ. These crystals are the only means of diagnosis, and they are very characteristic. Cystin is soluble in caustic ammonia, and may be precipitated by acetic acid, and crystallises in hexagonal tablets or plates. It also deposits from the urine spontaneously in similar forms.* This substance is of interest in a chemico-pathological point of view, from its containing so large an amount of sulphur as 27 per cent. In a clinical point of view its importance is chiefly in relation to the possible formation of a cystine calculus. The urine containing cystine when passed is usually feebly acid, has a peculiar odour, is slightly turbid, and deposits the characteristic crystals on cooling. It has been most frequently observed in young women of chlorotic aspect and condition. Dr. Roberts' work contains an excellent account of this singular disorder.

Leucin and tyrosin. These crystalline substances have been found in the urine, but they have no affinity to renal disease. They are significant of acute atrophy of the liver. They have also been detected in the urine of typhus patients. Dr. Roberts states that the urine of a patient suffering from acute yellow atrophy of the liver, after standing forty-eight hours, deposited an abundant sediment of tyrosin, crystallised in sheaf-

* The urine in which cystin occurs has a colour of an unripe lemon, greenish yellow.

CHEMICAL
PROPERTY.

Leucin and tyrosin. like bundles of acicular crystals.* Leucine occurs in more rounded masses. Dr. Thudichum calls them balls or rhombic plates. In Otto Funke's plates they appear as fine needle-like lines, or aggregated in a stellar form; the radii either infinitely acute, or presenting rhombic prismatic terminations.

Creatin and Creatinine. These are usually included among the substances known as extractives in the analysis of urine. They occupy, apparently, a transitory position between the metamorphosis of muscular tissue and the ultimate development of urea and uric acid. They are of physiological and chemical interest, but of no pathological significance, except that recent investigations on the theory and subject of uræmia seem to prove that the symptoms of what is called uræmic poisoning, and attributed by Frerichs to the decomposition of urea in the blood, and its conversion into carbonate of ammonia, arise essentially from the accumulation in the blood of these products of tissue metamorphosis, creatin, creatinine, sarcine, and other extractives, which should be further transposed into urea and uric acid before excretion by the kidneys. , (Dr. Roberts, p. 359.)

Bile pigment. In jaundice, and in several organic diseases of the liver, the bile pigment finds its way into the urine.

Cholepyrrhin. Bilifulvin. Biliverdin. The brown pigment of fresh bile, on exposure to the air, rapidly passes into a greenish colour—the biliverdin.
To determine the presence of bile pigment in the urine, a thin layer of urine is poured on the back of a white plate, or on a white tile, and to the

* Loc. cit., p. 62.

CHEMICAL PROPERTY.

Bile pigment. centre of which a drop of commercial nitric acid is let fall, when, if bile pigment be present, an iridescent play of colour takes place, commencing with a greenish blue, and passing into violet, and then red, to a rosy hue, which soon fades into yellow, the bile pigment being eventually decomposed.

The method of testing may be varied by half filling a test-tube with urine, and carefully pouring down the side of the tube a small quantity of nitric acid, taking care that the fluids do not mix. The acid being the heavier fluid will pass to the bottom, and at the junction of the urine with the acid, layers of the above colours will be displayed in the order there described.

If a small quantity of chloroform be shaken up in a tube with urine containing bile pigment, the colouring matter is extracted from the other ingredients, and remains in solution in the chloroform. If this be evaporated carefully, the pigment left, gives, on the addition of a drop of nitric acid, a beautiful ruby red colour.

This method by chloroform is a very delicate test, as the slightest trace of bile pigment is taken up by the chloroform, and the application of nitric acid will, even with very minute traces, exhibit the characteristic sequence of colours. This method is also readily available for discovering bile pigment in other fluids—the serum of dropsical cavities, &c.

Cholic acid. Lehmann says, the presence of the biliary acids in the urine is not so rare as has generally been supposed. Pettenkofer has detected cholic acid in the urine in a case of pneumonia. In most cases of jaundice from simple obstruction in the bile ducts, however abundant the pigment in the urine may be, they are absent; while in other cases with very

CHEMICAL PROPERTY.

Cholic acid.

little pigment the biliary acids have been found. It is thought that their presence in the urine indicates organic disease of the liver.

Pettenkofer's test.—The following method is recommended for the detection of cholic acid in the urine. A certain bulk of urine is evaporated to a syrup, and then extracted with pure alcohol. This alcoholic extract is moistened with a little water, to which is added a few drops of a solution of sugar of the strength of one part of sugar to four of water. Pure sulphuric acid, free from sulphurous acid, is then added cautiously by drops to the mixture, which then becomes turbid; but on carefully adding drop by drop fresh portions of sulphuric acid, the fluid becomes clear, at first yellow, rapidly passing into cherry red, then to deep carmine, and thence to purple, and finally into a deep violet blue. Too much solution of sugar frustrates the test; and care is requisite to prevent too rapid rise of temperature on addition of the sulphuric acid.

Chemists are, however, not agreed as to the certainty or value of the test. According to a recent edition and translation of Neubauer and Vogel's 'Analysis of the Urine,' oleic acid and albumen give analogous reactions.*

* See French edition. Paris, 1870. Page 105.

Morphological Constituents of the Urine.

ORGANIC FORMS.

Mucus in the Urine.

The protective epithelial cells of the urinary passages are, in minute quantity, constantly present in healthy urine. When they do not exceed the healthy proportion, they are seen as a very delicate haze occupying the inferior layer of urine when set at rest in a conical urine glass. It is always more abundant in the urine of women than of men. This will, according to the sex of the patient, exhibit to the microscope urethral or vaginal squamous epithelium, or vesical mucous corpuscles. The latter usually predominate in the urine of men, while in women there may be a notable amount of squamous epithelium derived from the vagina; and this is especially abundant in those who suffer from leucorrhœa. The student should familiarise himself with these several forms, that he may learn to distinguish between natural epithelium and those modifications of it that are significant of disease.

An increase in the apparent amount of mucous corpuscles denotes some disturbance in the urinary outlets. In old standing cases of urethral stricture mucous corpuscles are abundant. A common cause for their increase is an excessive acidity of the urine. They are abundant in urines containing crystals of uric acid and oxalate of lime. They are much augmented in gravel or lithiasis, and in gout. They gradually increase in quantity after the stage of hæmaturia in calculous disease of the kidney.

In some dyspeptics, quite apart from renal irritation, from a nervous irritable state of the system

ORGANIC
FORMS.

Mucus in the Urine. generally, the mucous corpuscles are wont to be in greater proportion than in health. In cystitis from whatever cause, catarrh of the bladder, gouty or calculous cystitis, the amount of mucus passed with the urine is very great; but this is usually accompanied by pus-cells, and those organic products rapidly induce decomposition of the urea, and the mucous and pus-cells become disintegrated or broken up in part by the volatile alkali, and a viscid magma or jelly is formed immediately after the urine is passed, or even while retained within the bladder.

Pus in the Urine. There is a natural transition from the mucus-corpuscle to the pus-corpuscle.

Mucus-corpuscles in abundance are the forerunners of the pus-corpuscle from all epithelial textures. It is not possible to distinguish between the two cells in the transition stage. The mucus-cell is spherical and mono-nuclear. The earliest of the pus-cells are also spherical and mono-nuclear. But the nucleus is seen to elongate, to become reniform, and soon trefoiled, or with a triple nucleus, and occasionally as many as four may be seen. Dilute acetic acid added to these cells render the nuclei more visible. The fluid (liquor puris) accompanying these cells is albuminous, so that purulent urine always displays this property.

Pus derived from the kidneys, or the pelvis of the organ, is perfectly miscible with the urine; consequently the urine, when passed, is slightly opaque or creamy, and, set at rest in a conical urine glass, separates into two distinct parts—a clear upper portion, and a distinct sediment, yellow or creamy in colour. The pus-corpuscles, falling as a precipitate by their own weight, settle at the bottom with a well-defined surface separating them from the

ORGANIC FORMS.

Pus in the Urine.

urine above. Purulent urine has a faint acid reaction on being passed, but very rapidly undergoes decomposition on exposure to the air. Pus is present in the urine in the following disorders:

In nephritis; in calculous nephritis; in tubercular nephritis; in calculous pyelitis; in tubercular pyelitis.

Pus-cells may be seen in other disorders, but they are either few in comparison with these diseases, or the urine is viscous from admixture of mucus, as in diseases of the bladder or prostate.

In gonorrhœa, gleet, and stricture, either mucous or pus-cells may be visible, but they are never otherwise than accompanied by shreddy membranous-looking particles, or the pus-cells are aggregated together in groups or films. In leucorrhœa pus-cells may be occasionally present, but they are accompanied by an abundance of the squamous epithelium of the vagina and os uteri.

The epithelial gland-cell from the renal tubes is never seen in healthy urine.

Blood in the Urine.

Blood-corpuscles in the urine are always seen isolated or distinct, never in rouleaux, as in blood from other sources. They may be entangled in fibrinous or membranous shreds, as seen in the hæmaturia significant of renal calculus. Or they may be formed into casts, as in the hæmaturia of scarlatinal dropsy and acute morbus Brightii. (See blood-casts.) Blood in the urine, from whatever part of the renal or urinary apparatus, will always give an albuminous quality to the urine.

Diseases in which blood appears in the urine:— Idiopathic nephritis; traumatic nephritis from external injury; blows; penetrating wounds, &c.

ORGANIC FORMS.

Blood in the Urine.

Turpentine, cantharides, taken internally or applied to the skin (rarely).

First stage of acute albuminuria, after scarlet fever, in acute morbus Brightii:—Gouty nephritis; lithiasis or gravel; calculous nephritis and pyelitis; tubercular nephritis; cancer of the kidney; endemic hæmaturia from presence of Bilharzia hæmatobia; prostatic disease; calculus or stone in the bladder; polypus of the bladder; in purpura hæmorrhagica; occasionally in severe gonorrhœa.

Renal casts.

RENAL CASTS, *and their significance.*—In describing the symptoms of the several forms of acute and chronic morbus Brightii, the casts which are washed from the uriniferous tubes were mentioned as characteristic either of a particular form or of a given stage of the disease.

I add here a special list of all the forms of casts with the disease or stage of progress of which it is significant.

Blood-casts.

Fibrinous casts.

The blood-casts and the fibrinous blood-casts are synonymous. They are fibrinous moulds of the tubes containing blood-corpuscles. They arise from blood escaping from the Malphighian bodies and coagulating in the tubes. They occur in the early stage of scarlatinal dropsy and of acute morbus Brightii. In the hæmaturia produced by turpentine or cantharides; in the hæmaturia from injury by wound or otherwise of the kidneys, but not in the hæmaturia of calculous disease of the kidney, nor in tubercle, nor in cancer.

The epithelial cast.

After the congestive or hæmorrhagic period of acute albuminuria has passed the epithelial gland-cells of the uriniferous tubes are cast off in abund-

ORGANIC FORMS.

The epithelial cast. ance; they are washed out by the current, and appear in the sediment of the urine as moulds of the tubes made up of these cells held together by some fine granular matter. When the cells are most numerous it is called an epithelial cast. When the granular matter predominates and separates the cells individually from each other, it is called a granular epithelial cast. In the transition period of the acute form blood-corpuscles often accompany the epithelial cell.

With these epithelial casts there are constantly seen, both in the cast as well as free or isolated, cells of a size much larger than the epithelial cell; these are not only highly granular, but contain numerous highly refractive nuclei. These are the large compound granule-cell, or Gluges' corpuscle.

These epithelial granule-casts with the large compound granule-cell are significant of the second stage of acute albuminuria, and they represent the acute desquamative nephritis of Dr. George Johnson, and the tubal nephritis of Dr. Dickinson.

Pus-casts. It occasionally happens that the epithelial cell is replaced by the presence of pus-corpuscles, which are moulded together as the epithelial cells are and washed out, not as isolated cells, but in these cylindrical groups. They are rare, but the cases do not differ, as regards symptoms, from those in which the epithelial cast alone is visible. They represent an acute form of albuminuria.

Fatty and granular casts. In the chronic varieties of albuminuria casts are seen free from cells, or the vestige only of one here and there; but these casts have a distinct granular appearance, and in the fine granular matter, or in the more transparent part of the cast, numerous

ORGANIC FORMS.

Fatty and granular casts. isolated, highly refractive grains or molecules are seen.

These appear to be the nuclei of abortive or of broken-up disintegrated cells. When large, or a few of them coalesce together, they have all the characteristic refraction of fat-grains. These when grouped together in granules, both of large and small dimension, give a distinctly fatty aspect to the cast, and these are designated fatty casts. They are most characteristic of the fatty and amyloid form of degeneration.

Granular casts are also met with in the chronic albuminuria of cardiac and pulmonary dropsy. In these dropsies with albuminuria casts are occasionally seen which are perfectly hyaline or transparent, with a well-defined outline, but with fissures as if the mould were split.

Casts very densely granular are met with in the amyloid form of disease, and Dr. Dickinson says they will receive the characteristic orange-brown stain by aqueous solution of iodine.

Hyaline casts. These were formerly very inappropriately termed glassy casts. They are perfectly transparent moulds or cylinders, with a well-defined outline; in their simplest form they are free from all other material, perfectly transparent and colourless. Mostly, however, they have here and there a minute isolated granule, or sometimes even a compound granule-cell attached to them. Sometimes they present a remarkable fissure or split half across the diameter.

They are met with in the latter stage of the acute forms of albuminuria, and when they succeed to the epithelial casts of the earlier period, they are of favorable import—representing what may be termed

ORGANIC FORMS.

Fatty and granular casts. a catarrhal state of the tubes, and implying a gradual subsidence of the graver symptoms. They are also met with in all the chronic forms of albuminuria, when their significance will depend greatly on the accompanying symptoms of the patient. In every case they represent a passive, not an active state of disease.

These casts have their analogue in the pituitary secretion which is produced in catarrhal conditions of the bronchial tubes.

Tubercular matter. From the several cases of tubercular disease of the kidney which I have had an opportunity of examining, I greatly doubt if there be any conditions of the pus and urine, which in themselves are pathognomonic of tubercle. The diagnosis must rest on the antecedent and concurrent symptoms. It has been thought that the presence of granular matter in the pus, insoluble in dilute acid, was symptomatic. I have not been able to verify this statement.

Cancer cells. In rare cases these may be seen in the urine. Large fusiform or pyriform cells containing a variety of nuclei represent their general character.

Spermatozoa. *Spermatozoa.* — The only importance of the presence of the spermatic germs in the urine, in relation to diseases of the kidney, rests on the fact that urine containing a minute quantity of the spermatic fluid will yield evidence of a trace of albumen. It happened recently to me to see a gentleman very nervous about his health, and who had got into the foolish habit of testing his own urine, and who believed that he was likely to have

ORGANIC FORMS.

Sperma-tozoa. albuminuria, often bringing me samples of his water in which he thought he had detected albumen, but in which none could ever be detected by myself, did on one occasion bring some passed that morning before rising which he said was unmistakeably albuminous, although only slightly. Nitric acid and heat, and nitric acid alone, as well as tincture of galls, gave decided proof of the presence of a trace of albumen. On examination by the microscope a considerable number of spermatic germs were seen, which at once proclaimed the source of the albumen, and convinced the patient that the cause of his anxiety was found in the ordinary healthy function of the organs of generation, for that morning an act of coition had immediately preceded the last micturition, and once more the nervous fear of impending morbus Brightii was dispelled.

Vibrios. These make their appearance only on the commencement of the putrefactive process. They are significant only of the decomposition of the urine.

Sporules. In urine containing sugar or albumen, or both, at a temperature not lower than 70° Fah., and by free exposure to the air, a number of minute bodies will be generated, which are either the sporules of the penicilium glaucum, or those of the sugar fungus—the torulæ cerevisiæ. Dr. Hassall communicated to the Medical and Chirurgical Society an important paper, which was published in the 36th volume of the 'Transactions,' 1853, "On the Development of Torulæ in the Urine," in which he demonstrated very clearly the distinction between the sporules developed in albuminous, and those in saccharine urine. These sporules are developed in

ORGANIC FORMS.

Sporules. urine in which the quantity of sugar is so minute that the copper test fails to detect it; and, therefore, the formation of those torulæ or germs has been suggested as a test for sugar in the urine when the quantity may be too minute for the chemical agency to determine its presence. The sporules of the penicilium glaucum, arising in albuminous urine, are very easily distinguished from those of the torulæ cerevisiæ. Dr. Hassall's paper is illustrated with drawings tracing these growths from the sporule stage to that of the filament and thallus, and ultimately to that of complete fructification. The distinction between the sporules of albuminous urine and those of saccharine urine are of importance to the student. They may both occur in the same urine, however. The sporules of penicilium glaucum are small, minute, ovoid spores, which soon elongate, and grow by lengthening in the long diameter, and presenting slowly appearances similar to a string of sausages. The sporule of the sugar fungus is much larger, spheroidal, has a double wall, and the contents of the cell are granular and nuclear; growth also takes place by elongation, but the filamentous stage is more distinctly bead-like than that of the penicilium. The student should develop these sporules in urine and study the forms for himself.

Ova.   *Ova.*—Reference has already been made to the endemic hæmaturia of certain countries—the Cape of Good Hope, Egypt, &c.

From the careful observations of Dr. John Harley we are in possession of facts connected with the development of the ova of the hæmatoid worm—the Bilharzia hæmatobia. In a paper published in the 47th volume of the 'Medico-Chir. Transactions,'

ORGANIC FORMS.

Ova. the ova of this entozoon are figured. The student is referred to these plates, and the accompanying paper, for a succinct history of these entozoa.

Echino- *Echinococcus in the urine.* — Hydatids in the
cocci. kidneys are of rare occurrence. Some two years since a man who was occasionally employed by me in the country as a plumber and painter, and who had fallen into ill health, incidentally mentioned to me that he was troubled in his water; and on bringing to me a sample, I found it was filled with hydatid cysts, varying in size from a millet seed to that of a good-sized pea. He had been subject to epileptiform seizures from his youth. Latterly he suffered much from spasmodic asthma. Eventually he became dropsical and died. But to the last he passed immense numbers of these echinococci. The urine accompanying them was from the first examination of it albuminous.

No post-mortem examination was made.

THE END.

INDEX.

A.

	PAGE
Acidity of the urine	181
Acute albuminuria	18
—— morbus Brightii	22
—— —— morbid anatomy of	24
—— —— treatment of	75
Albumen in the urine	213
—— temporary or permanent	214
—— process for estimating	214
Albuminose, modified albumen	216
Alcohol in urine	211
—— test for	212
Alkalinity of urine	182
—— causes and effects of	183
Amyloid disease, nature of	137
—— Dr. Dickinson's researches on	137
—— tests for	139
—— symptoms of	159
—— treatment	160
Anstie, Dr., on alcohol in urine	212
Arteries, renal, changes of structure of	146
Atrophic nodular kidney	161

B.

Beale, Dr. L., on chlorides in urine	209
Bile pigment in urine, tests for	225
Bird, Dr. G., on oxaluria	187
Blood in urine	229
—— casts	229
Bright's disease, acute. *See* Albuminuria	22
—— —— morbid anatomy of, 1st, 2nd, 3rd stages of	129
—— chronic	133
—— —— forms of	134
Bristowe, Dr., on renal cysts	165

C.

	PAGE
Calculous nephritis	30
Cantharidis, action on the kidneys	11
Cancer cells in urine	232
Casts in urine, various forms of	229
Chemical properties of urine	181
Chlorides (fixed) in the urine	209
Cholic acid in urine, tests for	225
Chronic albuminuria, cause and pathology	112
—— symptoms	122
—— gout, symptoms of	123
—— after scarlet fever	119
Chylous urine	183
Cirrhosis of the kidney, Dr. George Harley on	143
Colour of the urine	172
—— varieties of	174
Contracted red kidney	133
Creatin and creatinine, their significance	226
Cysts, renal	163
Cystic degeneration, Dr. Bristowe on	165
Cystin in urine	224

D.

Dealkalized fibrine	138
—— how formed	138
—— Dr. Dickinson on	138
Diabetes, urine in, aromatic	177
Dickinson, Dr., on amyloid	116
Diphtheria a cause of albuminuria	19
Dupré, Dr., on alcohol in urine	211

E.

Echinococci in the urine	234
Embolism of renal arteries	26
Endemic hæmaturia	47
—— nephritis, symptoms	107
Epithelial casts	229
Erythrine	172

F.

Fatty and granular casts	230
Fehling's solution for estimation of sugar	220

	PAGE
Fibrinous casts	229
—— blood-casts	229
—— coagula in renal arteries	26
Fibrine dealkalized, how formed	138

G.

Gout and lead poisoning	14
Gouty nephritis	27
—— kidney	141
—— atrophic nodular kidney, in	161
—— nephralgia	230
Granular casts	230

H.

Hæmaturia in gouty nephritis	28
—— endemic nephritis	47
Harley, Dr. Geo., on cirrhotic kidney	143
Heart disease, cause of albuminuria	21
Hippuric acid in urine	199
—— crystals in urine	200
Hyaline casts in urine	237

I.

Inflammatory dropsy	22
Intemperance, cause of both acute and chronic renal disease	125
Iodoform, test for alcohol	212

J.

Johnson, Dr. George, his researches	129

K.

Kidney, inflammation of. *See* Nephritis.	
—— cancer of	232
—— tubercle of	232
—— chronic disease of. *See* Chronic Albuminuria.	
—— contracted, small	133
—— —— symptoms and treatment of	157
—— large granular	136
—— large smooth	19
—— amyloid, disease of	137
—— atrophic gouty	141
—— —— symptoms and treatment of	162
Kystein	184

L.

	PAGE
Lead poison predisposes to renal disease	14
Lehmann on oxalic acid	189
Leucin and tyrosin in urine	225
Lieben, M., process for detecting alcohol in urine	213

M.

Micturition, frequency of	30
—— a symptom of	180
Mucus in urine	226

N.

Nephralgia, gouty, symptoms	77
Nephritis, definition of	2
—— varieties of	4
—— acute	18, 22
—— chronic	112
—— cancerous	46
—— —— symptoms	102
—— tubercular	93
—— endemic	47
—— —— symptoms	107
—— and pregnancy	48
Neutrality of urine	182
Nitrate of potash, action of, on the kidneys	13

O.

Odour of the urine	171
—— fœtid	177
—— aromatic	177
—— ammoniacal	177
Ova in urine	224
Oxalate of lime, not present in fresh urine	193
Oxalic acid diathesis	196
—— source of	189
—— chemistry of	190
—— Dr. Owen Rees on	188
Oxaluric acid	192
Oxalurate of ammonia	191
—— M. Schunck on	192
Oxaluria, Dr. Prout and Dr. G. Bird on	187
—— symptoms of	196
—— significant of uric acid	187
—— treatment	196

P.

	PAGE
Peri-nephritis	49
—— symptoms	104
Pettenhofer's test for cholic acid	235
Phosphates in the urine	204
—— earthy and alkaline	204
—— —— determination of	208
—— of lime, crystalline form of	205
Physical properties of urine	171
Polariscope	223
Pregnancy and albuminuria	107
Prout, Dr., on oxalic acid diathesis	187
Pus in the urine	227
—— casts	230
Pyelo-nephritis from stricture	35
—— symptoms of	91
Pyo-nephrosis	108

R.

Rees, Dr. Owen, on oxalic acid diathesis	188
Renal colic	29
—— arteries, embolism of	26
—— —— fibrous coagula in	26
—— casts	229
Rheumatic nephritis	25
—— symptoms	76
Roberts, Dr., on crystalline phosphate of lime	205
—— his fermentation test	229

S.

Scarlet fever and acute albuminuria	64
—— and chronic albuminuria	112
Schunck, M., on oxalurate of ammonia	191
Scrofulous pyelitis	36
—— pus, obsolescence of	44
Simon, Mr., on renal cysts	164
Specific gravity of urine	178
Spermatozoa in urine	232
Sporules in urine	203
Stewart, Dr. Grainger, on amyloid	129
Stone in the kidney	30
—— pathology of	33
Sugar in urine	217
—— tests for	218

	PAGE
Sugar, quantitative determination by fermentation	219
Sulphates in urine, their significance	211

T.

Tubercular nephritis	36
———— symptoms and treatment of	98
———— pyelitis	36
———— matter in urine	232
Turpentine, action on kidneys	14

U.

Urates in urine	186
Urea	201
———— determination of	202
Uric acid, action of alkaline salts on	89
———— estimation of	185
———— in gout	186
———— oxidation of	189
Urine, clinical significance of	169
———— physical properties of	171
———— chemical properties of	181
———— organic forms in	226

V.

Vibriones in urine	233

W.

Weber, Dr. Herman, on chronic albuminuria after fever	119

ERRATA.

Page 12, line 12, *for* in which *read* for whom.
,, 19, ,, 23, *for* is seen studded *read* smooth or studded.
,, 33, ,, 7, *for* is formed *read* be formed.
,, 54, ,, 28, *for* discoloured *read* disordered.
,, 69, ,, 30, *for* rigour *read* rigor.
,, 78, ,, 3, *for* The urine *read* The uric acid.

www.ingramcontent.com/pod-product-compliance
Lightning Source LLC
Chambersburg PA
CBHW021345230426
43666CB00006B/415